# Software Testing

## Concepts and Practices



**K. Mustafa**
**R.A. Khan**

Alpha Science International Ltd.
Oxford, U.K.

**K. Mustafa**
College of Computer Science and Information Technology
King Faisal University
KSA

**R.A. Khan**
Department of Information Technology
School for Information Science and Technology
Babasaheb Bhimrao Ambedkar University
(A Central University)
Lucknow, India



Alpha Science International Ltd.
7200 The Quorum, Oxford Business Park North
Garsington Road, Oxford OX4 2JZ, U.K.

**www.alphasci.com**



ISBN 978-1-84265-367-8

Printed in India

# Software Testing

## Concepts and Practices

To my Grandmother
and
To my Father, **Naseer Khan**, for teaching me how to choose the best road to follow among life's many choices.

– R.A. Khan

# Preface

*Software Testing* presents very interesting anomaly for developers and testers. Bugs in critical systems have caused airplane crashes, allowed space shuttle missions to fail, halted trading on the stock market, and worse. Bugs can kill. Bugs can cause disasters. The so-called year 2000 (Y2K) bug has given birth to a cottage industry of consultants and programming tools dedicated to making sure the modern world doesn't come to a screeching halt on the first day of the next century. Debugging is a narrow view of software testing, which can be performed to find out design defects. Software testing activities play a critical role in the production of dependable systems, and account for a significant amount of resources including time, money, and personnel. The difficulty in software testing stems from the complexity of software. A program cannot be completely tested with a moderate complexity. Software testing is a critical part of the software process, which comprises about 40 percent of the effort spent on the software process. Therefore, it appears high time to introduce software-testing aspects to software engineers of today rather than wait for them to learn through experience, at a high cost.

The book has been planned and designed for a one-semester course on 'software testing' as a follow up course on software engineering. It is aimed to fit in the undergraduate programmer in computer science and master's program in computer applications being conducted by most of the higher educational institutions. This book points a pragmatic picture of software testing research and practices, and hence may be suitably adapted for undergraduate allied courses offering 'software testing and quality assurance' and for software engineering practitioners to enhance their effectiveness.

## KEY FEATURES

This text, focusing on software quality and its assurance in practice, has been planned to suit the needs of its audience. Key features that distinguish it from other books include the following:

- Broad Audience: Unlike other software testing books that are commonly designed by keeping software testing practitioners in mind, this book is prepared for a larger audience including academicians and practitioners.
- Software Testing Embedding: Concepts of software testing have been modeled as a phase-embedded activity rather than treating them as separate and post development activity.

- Learning by Objectives: Each chapter of the book starts with a set of objectives, to which a prospective should be targeted to achieve rather than leaving the student directionless.
- Review Questions: Each chapter ends with three types of review questions including objective, short answer, and descriptive thought provoking questions.
- Key Terms: Each chapter ends with the list of key terms, as these terms generally refer to certain abstract concepts, may be used for better and crisp communication.
- Solution Manual: The book is packaged with a solution manual in electronic form, which may be made available to qualified teachers on demand.
- Instructor's Manual: It includes PowerPoint slides, brief notes on teaching and list of projects with descriptions.
- Bibliography: Each chapter ends with a list of key references for the concepts in the chapter, to enable the users to find in-depth information related to the content of the chapter.
- Useful Links: Lists of useful website address have been appended to each of the chapters for quick references on needed topics.

## ORGANIZATION OF THE BOOK

The book is addressed to those with adequate knowledge of software development and engineering activities, and has been organized in eleven chapters as follows:

### *Chapter 1: Software Testing Fundamentals*

In this chapter we address testing in general and appreciate the needs, importance and significance of software testing in software development. We define the purpose of software testing, its properties, components and types. The role of software testing has been identified in software quality assurance. We have also tried to relate software testing with software reliability. Various facts with regard to software testing have been realized in this chapter. Numerous software testing characteristic have been enumerated and different perspective have been presented for defining software testing.

### *Chapter 2: Software Testability*

This chapter includes software testability as a key factor to software quality, and appreciates the need, importance and significance of software testability metrics. We also define testability estimation framework. The facts related to various features to software testability metrics have been realized. We also describe the software testability factors and their contribution.

### *Chapter 3: Static Testing*

This chapter includes an overview of strategy, appreciates the need, importance and significance of software testing strategy. In addition, we define strategic approach to software testing, by realizing the facts related to verification and validation process. We also visualize the different principles of static analysis, and understand the prescriptive process to perform walkthrough and inspection during software testing. At last we realize the importance and significance of technical review in software testing.

### *Chapter 4: Black Box Testing*

This chapter includes the descriptions on the benefits of using black box testing for the projects. We present advantages and disadvantages of black box testing technique, and appreciate the goals and responsibility of boundary value analysis. We have also defined boundary conditions. Robustness testing and its methodology have been discussed in this chapter. In addition, syntax testing and finite state testing has been included in this chapter.

### *Chapter 5: White Box Testing*

This chapter includes a description on white box testing techniques, importance of using basis path testing, and significant methodology. We further address on the role and purpose of control flow graph, identify the benefit of using graph matrix in software testing, and define control structure testing including condition coverage, branch coverage, and statement coverage. In addition, mutation testing has been introduced to test object oriented software.

### *Chapter 6: Software Testing Strategies*

This chapter focuses on strategy in general. We appreciate the need, importance and significance of software testing strategy. We further address on generic view of software testing strategy, and visualize the different perspectives of unit testing, integration testing, validation testing and system testing techniques. Criteria for completion testing have been identified in this chapter. At last, we realize the importance and significance of software component testing.

### *Chapter 7: Planning for Software Testing*

This chapter focuses on establishing software test plan in advance to test the software. We also realize the need and importance of documentation in test specifications. Furthermore, criteria for successful testing have been discussed in detail. It has been identified what to test or not to test. A list of deliverables has been presented then after.

***Chapter 8: Object Oriented Testing***

This chapter describes object oriented software testing in general and appreciates the need, importance and significance of object oriented technology. We also define and understand the software characteristics contributing to testing and quality. Furthermore, principles of testing object oriented software has been discussed in this chapter. We describe the major differences between object oriented paradigm and traditional paradigm.

***Chapter 9: Software Fault Tolerance***

This chapter describes fault tolerance in general and appreciates the need, importance and significance of software fault tolerance. We also define and understand software failure. This chapter realizes the facts related to various checks available for error detection, and enumerates numerous techniques of software fault tolerance. We also visualize the different perspectives of detecting software errors.

***Chapter 10: Design Phase Mutation***

In this chapter we address the need, importance and significance of introducing mutation during testing object oriented software early in the software development life cycle. We also define mutant operators in design phase. This chapter estimates adequacy of test data set, and realizes the importance of creating mutant operators in estimating software test adequacy.

***Chapter 11: Class Test Complexity***

This chapter focuses on the current scenario of object oriented metrics. It has been deduced that early availability and use of metrics in SDLC is a key factor to a successful management of software development, and a leading factor for reducing the class complexity test effort. A metric suit has been proposed and validated for predicting class test complexity of object oriented software. We further set a regression line correlating the software characteristics with complexity.

K. Mustafa and R.A. Khan

# Contents

## Chapter 1

# Software Testing Fundamentals

*It's better to know some of the questions than all of the answers.*
**James Thurber**

## OBJECTIVES

A successful completion of this chapter will enable the reader for following:

- To understand and describe software testing in general.
- To comprehend different historical definitions of software testing.
- To appreciate the need, importance and significance of software testing.
- To list purpose, goals and objectives of software testing, its properties.
- To relate and compare different types of testing.
- To identify the role of software testing in software quality improvement.
- To differentiate software testing from other similar activities.
- To realize various challenges with regard to software testing.
- To enumerate and realize numerous software testing characteristics.
- To visualize different perspectives of software testing.
- To understand and describe foundations of software testing.
- To define and understand various testing related terminologies.
- To enumerate and realize numerous software testing principles.
- To list various software testing myths, and related facts.
- To realize the importance and significance of effective testing.
- To define and understand software testing effort estimation.
- To appreciate the need, importance and significance of mapping debugging and testing, with software development life cycle.
- To understand and describe different perspective of bugs.
- To apply the overall knowledge for relating better conceptualization of fundamental of software testing.

## 1.1 INTRODUCTION

Development of large software products involves several activities. These activities need to be suitably coordinated to meet the desired requirements. Because of the fallibility of its human designers and its own abstract, complex nature, software development must be accompanied by quality assurance activities. It is not unusual for developers to spend 40 percent of the total project time on testing. Testing is often considered an expensive and uncontrollable process, which takes too much time, costs more than planned, and offers insufficient insight of the quality of the test process. Testing is a process of planning, preparing, executing and analyzing the difference between the actual status and the required status. Software testing is a process of verifying and validating software application or program so that it meets the business and technical requirements that guide design and development, and works as expected. Test planning and preparation activities emphasize the fact that testing should not be regarded as a process that can be started when the object to be tested is delivered. Test planning decides an important defect, which from the customer's perspective affects the usability and functionality of the application, by reviewing the requirements and design documents. A test process requires accurate planning and preparation phases before any measurement actions can be implemented. Testing cannot show the absence of faults but- it may definitely show their presence.

Several historical definitions of software testing are available in the literature. All these definitions are useful, but in different ways. Some of the definitions focus on what is done while testing, others focus on more general objectives like assessing quality, customer satisfaction, and expected results.

- In 1973, Hetzel explained that 'Testing is the process of establishing confidence that a program does what it is supposed to do', in his book *Program Test Methods*. This definition focuses on the customer satisfaction.
- In 1979, Glenford Myers explained, 'Testing is the process of executing a program or system with the intent of finding errors', in his classic book, *The Art of Software Testing*. At that time, Myer's definition was probably the best available and mirrored the thoughts of the day. Myer's definition focuses on the fact that testing occurred at the end of the software development life cycle and its main purpose was to find the errors.
- In *The Complete Guide to Software Testing,* Bill Hetzel (1983) stated, 'Testing is any activity aimed at evaluating an attribute of a program or system. Testing is the measurement of software quality'.

Software testing is the process of verifying that a system satisfies its specified requirements or identifying differences between expected and actual results. The computer society defines testing as:

> *A verification method that applies a controlled set of conditions and stimuli for the purpose of finding errors. This is the most desirable method of verifying the functional and performance requirements. Test results are*

*documented proof that requirements were met and can be repeated. The resulting data can be reviewed by all concerned for confirmation of capabilities.*

ANSI uses two different definitions in the official standards, both of which appear somewhat unsatisfactory and validation oriented.

*The process of operating a system or component under specified conditions, observing or recording the results, and making an evaluation of some aspect of the system or component (IEEE/ANSI, 1990 [Std 610.12-1990]).*

*The process of analyzing a software item to detect the difference between existing and required conditions (that is, bugs) and to evaluate the features of the software items (IEEE/ANSI, 1983 [Std 829-1983]).*

Nowadays, software is becoming large and complex. Looking at software deliverable, it is high time to look at it in totality and not limit to program. Software consists of program, data and documents, as depicted in figure 1.1.

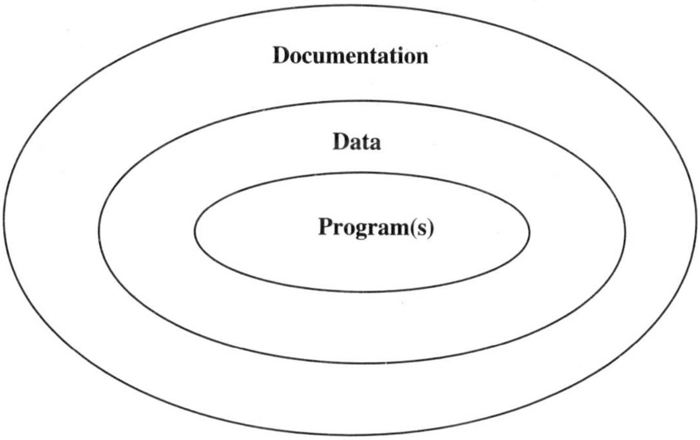


**Figure 1.1: Components of Software**

Thus seeing software testing in a limited manner, that is, program testing would be highly impractical. It appears that it is time to correct the vision of testing and broaden the scope of all the components of software. Software testing is an integral part of the system development function, not an afterthought. It holds several characteristics including the following:

- Software testing may start with the requirements not the code.
- Software testing is both the static and dynamic activity.
- Software testing prevents the occurrence of failure.
- Software testing helps in finding the faults sooner in the development life cycle, and hence reduces the cost to fix the bugs.
- Software testing process creates re-usable testwares.

Software testing is a group activity and not a one-person job. The size of the team varies. The size of the team may be larger or smaller depending on the size and complexity of the application being tested. The programmer(s) who wrote the application may have a reduced role in the testing process after completion because they may not be able to take an unbiased look at the results of their labors. Testers must be cautious, curious, critical but non-judgmental, and good communicators.

## 1.2 SOFTWARE TESTING PERSPECTIVE

Software testing is often a very costly and generally under-honored component of its life cycle. Software development processes typically focus on avoiding errors, detecting and correcting software faults that do occur, and predicting reliability after development. It is believed that software industry is at a risk for a disaster of some kind, a disaster in which the blame will clearly lay on the software. Many computer systems are used in critical applications such as spacecraft and defense systems. When lives and fortunes depend on software, software quality and its verification demand increased attention. Increasing emphasis placed on high quality and customer satisfaction of software calls for rethinking on the objectives and management of testing. Software testing has an obvious role in finding bugs, and less obvious role in evaluating reliability. Test and evaluation methods and tools, in themselves, do not guarantee effective testing and ensure high quality of software. As the complexity of applications and the software increases, software testing and evaluation becomes more difficult and its effectiveness falls below expectations. Software testing is not an exact science; it is both an art and a science. But testing has often been pursued purely on technical grounds and during the past two decades there have been considerable advances in software testing techniques and methodologies, test case generation, CASE tools, etc.

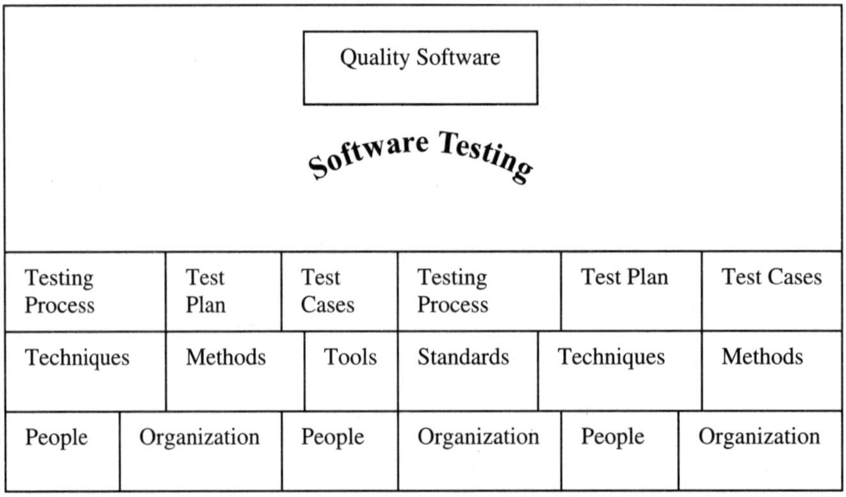


**Figure 1.2: Foundations of Software Testing**

The foundations of software testing are:

- Test process, test cases and test plan.
- Techniques, methodologies, tools and standards.
- The people and the organization

The test process, techniques and tools are significant contributors to effective and efficient testing and quality assurance. They can, however, offer better results only when they are built upon 'people foundation' and sound managerial and organizational culture. It is the people and the culture of the organization that determines how any system is practiced. Figure 1.2 shows the building blocks of software testing.

### 1.2.1 Related Terminology

Software testing has an established vocabulary by now. Brief descriptions of the related technical terms used in software testing process literature are as follows:

**Audit:** An independent examination of a work product or set of work products to assess compliance with specifications, standards, contractual agreements, or other criteria.

**Boundary Value:** A data value that corresponds to a minimum or maximum input, internal, or output value specified for a system or component.

**Boundary Value Analysis:** A selection technique in which test data are chosen to lie along boundaries of the input domain [or output range] classes, data structures, procedure parameters, etc. Choices often include maximum, minimum, and trivial values or parameters.

**Branch Coverage:** A test coverage criterion, which requires that for each decision point each possible branch be executed at least once.

**Bug:** A fault in a program, which causes the program to perform in an unintended or unanticipated manner.

**Coverage Analysis:** Determining and assessing measures associated with the invocation of program structural elements to determine the adequacy of a test run. Coverage analysis is useful when attempting to execute each statement, branch, path, or iterative structure in a program.

**Crash:** The sudden and complete failure of a computer system or component.

**Criticality:** The degree of impact that a requirement, module, error, fault, failure, or other item has on the development or operation of a system.

**Cyclomatic Complexity:** The number of independent paths through a program, is equivalent to the number of decision statements plus 1.

**Error:** A discrepancy between a computed, observed, or measured value or condition and the true, specified, or theoretically correct value or condition.

**Error Guessing:** A test data selection technique, where selection criterion is to pick values that seem likely to cause errors.

**Error Seeding:** The process of intentionally adding known faults to those already in a computer program for the purpose of monitoring the rate of detection and removal, and estimating the number of faults remaining in the program. Contrast with mutation analysis.

**Exception:** An event that causes suspension of normal program execution. Types include addressing exception, data exception, operation exception, overflow exception, protection exception, and underflow exception.

**Exhaustive Testing:** Executing the program with all possible combinations of values for program variables. This type of testing is feasible only for small, simple programs.

**Failure:** The inability of a system or component to perform its required functions within specified performance requirements.

**Fault:** An incorrect step, process, or data definition in a computer program that causes the program to perform in an unintended or unanticipated manner.

**Oracle:** Any means used to predict the outcome of a test.

**Risk:** A measure of the probability and severity of undesired effects.

**Risk Assessment:** A comprehensive evaluation of the risk and its associated impact.

**Test:** A system or component execution under specified conditions to observe record and evaluate some aspect of the system or its component.

**Testability:** The degree to which a system or component facilitates the establishment of test criteria and the performance of tests to determine whether those criteria have been met.

**Test Case:** Properly specified inputs, predicted results, and a set of execution conditions for a test item.

**Test Case Generator:** A software tool, that accepts source code as input, test criteria, specifications or data structure definitions; uses these inputs to generate test input data; and, sometimes, determines expected results.

**Test Design:** Process of specifying the details of the test approach for a software feature or combination of software features and identifying the associated tests.

**Test Documentation:** Documentation describing plans for, or results of, the testing of a system or component. Its types include test case specification, test incident report, test log, test plan, test procedure, and test report.

**Test Driver:** A software module used to invoke a module under test and, often, provide test inputs, control and monitor execution, and report test results.

**Test Incident Report:** A document reporting on any event that occurs during testing requiring further investigation.

**Test Item:** A software component or module, which is the object of testing.

**Test Log:** A chronological record of all relevant details about the execution of a test.

**Test Phase:** The period of time in the software development life cycle in which the components of a software product are evaluated and integrated, and the software product is evaluated to determine whether or not requirements have been satisfied.

**Test Plan:** Documentation specifying the scope, approach, resources, and sc intended testing activities. It identifies test items, the features to be tested, the test responsibilities, required, resources, and any risks requiring contingency planning.

**Test Procedure:** A formal document developed from a test plan that presents detailed instructions for the setup, operation, and evaluation of the results for each defined test.

**Test Report:** A document describing the conduct and results of the testing carried out for a system or system component.

**Test Result Analyzer:** A software tool used to test output data reduction, formatting, and printing.

**Test Suite:** A set of one or more tests, usually aimed at a single object, with a common purpose and database, usually run as a set.

**Test Script:** A document, program, or object that specifies for every test and subtest in a test suite: object to be tested, requirement (usually a case), initial state, inputs, expected outcome, and validation criteria.

**Traceability Matrix:** A matrix that records the relationship between two or more products; for example, a matrix that records the relationship between the requirements and the design of a given software component. Closely related concepts include traceability and traceability analysis.

### 1.2.2 Myths

Various misconceptions about the software testing, its execution, usages, types and characteristics have been the point of discussion from the last several years. Few of them have been identified and discussed as myths along with respective realities:
as input

***Myth#1:*** ***Testing gets in the way.***

√ *Testing is as important as any other activity in software development/maintenance, and hence it should get due weightage. It can be complementary as well as integral part of software development life cycle.*

***Myth#2:*** ***The idea of testing to find faults is fundamentally wrong.***

√ *Testing facilitates to uncover errors and faults. Knowledge of errors and faults leads to its eradication. So an intentional effort, rather than incidental, is highly solicited to assure quality of software and cost-effectiveness.*

***Myth#3:*** ***Testing is destructive and late process; it is not a good use of creative abilities and technical skills.***

√ *Testing is a diagnostic and highly creative activity, which promotes quality. It has emerged as highly technical and skilled task, along with growth in size/ complexity/versatility of software.*

***Myth#4:*** ***Testing is a structured waterfall idea.***

√ *Testing may be purely independent, non-obstructing, incremental and iterative activity. Much of its nature depends upon the context and adapted strategy.*

***Myth#5:*** ***Testing is trivial.***

√ *Adequate testing requires a thorough understanding of system under test, appreciate strategy and testing techniques. Moreover, a foolproof mechanism of implementation is highly desirable.*

***Myth#6:*** ***Testing is simply poking around until you run out of time.***

√ *Good process of testing assures uncovering of maximum number of errors, with least amount of effort and time. Such a task is highly planned, structured and managed using appropriate strategy, techniques and tools.*

***Myth#7:*** ***Automated GUI testing is sufficient.***

√ *GUI commands supplied by command playback tool may not sufficiently test but do it partially.*

***Myth#8:*** ***If programmers are more careful, testing is unnecessary.***

√ *Programming errors can be minimized, but may not be eliminated being a human activity. Even the good programs are reported by studies to have 1-3 errors. Testing would always be a necessary component of software development and maintenance activities to assure correctness. Integration errors are highly likely many a bugs surface only during integration.*

***Myth#9:*** ***Testing is too expensive; we don't have time.***

√ *It stands true in case of testing 'Pay me now or pay me much more later'. There is always a cost associated with any good/fruitful activity, and so is with testing. The cost of finding and correcting errors is always higher as the time between fault injection and detection increases. Well planned and managed testing may prove to be very cheap.*

***Myth#10:*** ***Conventional testing is useless of objects.***

√ *Conventional testing techniques are highly independent of paradigm and context. These techniques may continue to apply in object- oriented software and other software applications. Moreover, with the advent of useful CASE tools it may prove to be easily applicable, more effective and efficient may be with some necessary changes.*

### 1.2.3 Purpose, Goal and Objectives

Software testing is a process of verifying and validating that a software application or program meets the business and technical requirements. Most general purpose of software testing is to achieve confidence that the software will work error free and correctly. Testing verifies whether the software meets its technical specification. Debugging is a narrow view of software testing, which can be performed heavily to program to find out design defects. The major objective of debugging is to finding the problems and get them fixed in the programming phase. The major objectives or goals of software testing may be described in five forms- verification, validation, defect prevention, quality improvement, and reliability estimation.

**Verification:** The verification process confirms that the software meets its technical specifications. Software verification examines the products of each development activity to determine if the software development outputs meet the requirements established at the beginning of the activity. A specification is a description of a function in terms of a measurable output value given a specific input value under specific preconditions. There are two types of verifications, that is, static verification and dynamic verifications. Software inspection, which concerns with analysis of the static system representation to discover problems, comes under static verification. Software testing, which concerns with exercising and observing product behavior, comes in dynamic verification.

IEEE defines verification as 'Confirmation by examination and provisions of objective evidence that specified requirements have been fulfilled'. There are four levels of verification:

*Component Testing:* Testing conducted to verify the implementation of the design for one software element (unit, module) or a collection of software elements.

*Integration Testing:* An orderly progression of testing in which various software elements and/or hardware elements are integrated together and tested. This testing proceeds until the entire system has been integrated.

*System Testing:* The process of testing an integrated hardware and software system to verify that the system meets its specified requirements.

*Acceptance Testing:* Formal testing conducted to determine whether or not a system satisfies its acceptance criteria and to enable the customer to determine whether or not to accept the system.

**Validation:** The validation process confirms that the software meets the business requirements. In practice, software validation activities may occur during software development life cycle to ensure that all requirements have been fulfilled. Since software is usually part of a larger hardware system, the validation of software typically includes evidence that all software requirements have been implemented correctly and completely, and is traceable to system requirements. A conclusion that software is validated is highly dependent upon comprehensive software testing, inspections, analyses, and other verification tasks performed at each stage of the software development life cycle. Tests with the purpose of validating the product works are named clean tests, or positive tests. Testing of device software functionality in a simulated use environment, and user site testing are typically included as components of an overall design validation program for a software automated device. A finite number of tests cannot validate that the software works for all situations. On the contrary, only one failed test is sufficient to show that the software does not work. A dirty test, or negative tests, refers to the tests aiming at breaking the software, or showing that it does not work. A piece of software must have sufficient exception handling capabilities to survive a significant level of dirty tests.

**Defect Prevention:** Defects are defined to be a variance between the expected result and actual result, when software is put to operation. The defect's ultimate source may be traced to a fault introduced in the specification, design, or coding phases. Early defect detection and prevention with static analysis is a key part of development process. Software defects increase the cost of developing software, increase the likelihood of annoying or losing customers, and increase the potential of lost revenue. The challenge for organizations that build or maintain large applications is to identify defects as early as possible in the development cycle. A defect that is identified and corrected while the developer is writing the code has a cost that is of magnitude lower than correcting the same defect in any of your existing test phases or once the product is released.

**Quality Improvement:** The demand for quality software continues to increase due to our society's increasing dependence on software and the often-devastating effect that a software error can have in terms of life, financial loss, or time delays. Embedded software system systems must ensure the consistent and error free operation every time it is used. The demand for increased software quality has resulted in quality being more of a differentiator between products than it ever has been before. Quality means the conformance to the specified design requirement. Being correct, the minimum requirement of quality, means performing as required under specified circumstances. One of the major factors driving any production discipline is quality. The software engineering approach works toward a single goal: to produce high quality software.

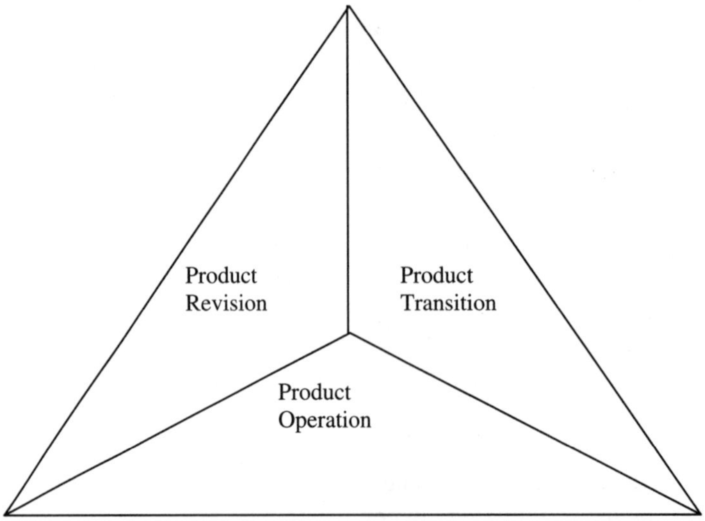


**Figure 1.2.3: McCall's Quality Factors**

As software is used in critical applications, the outcome of a bug can be severe, and hence can cause huge losses. Bugs in critical systems have caused airplane crashes, space shuttle missions to fail, halted trading on the stock market, and worse. Bugs can kill. The so-called year 2000 (Y2K) bug has given birth to a cottage industry of consultants and programming tools dedicated to making sure the modern world doesn't come to a screeching halt on the first day of the next century. In a computerized embedded world, the quality and reliability of software is a matter of life and death.

Software quality cannot be tested directly; the related factors that make quality visible can only be tested. J. P. Cavano and J. A. McCall have produced a quality view of a product as having three dimensions: product operation, product transition, product revision as shown in figure 1.2.3. The first factor, product operation, deals with quality factors such as correctness, reliability, and efficiency. Product transition deals with quality factors like portability and interpretability. Product revision is concerned with those aspects related to modifications of programs, including factors such as maintainability, and testability. Quality can be characterized by a series of properties referred to as the 'ilities'. These quality properties cover different aspects of quality of software system. The 'ilities' include: efficiency, flexibility, improvability, predictability, repeatability, robustness, sustainability, visibility, portability, reusability, usability, testability, understandability, adaptability, integrity, correctness, reliability, security, availability, scalability, maintainability, and time to market.

**Reliability Estimation:** A series of accidents caused by unreliable software proves that the reliability of software is a matter of life and death. Software reliability is defined as the probability of failure free operations for a specified period of time in a specific environment. No doubt, software reliability has important relations with many aspects of software, including the structure, and the amount of testing it has been subjected to. In the development stage of a software product, its reliability can be estimated by analyzing data obtained from software tests. Software testing can be costly, but not testing software is even more expensive, especially in places where human lives are at stake. Solving the software-testing problem is no easier than solving the halting problem. It can never be assured that a piece of software is correct. One of the major tasks in testing is to select the minimum set of test cases sufficiently effective for revealing potential faults in a program. But the fundamental question arises how? The test strategies are faced with the problematic question like this, since exhaustive testing is usually not tractable. It is evident that the evaluation criterion for test strategies may be used to measure the effectiveness of generated test cases. Mutation analysis is one of the evaluation techniques that assess the quality of test cases by examining whether they can reveal certain types of faults. It is sometimes helpful to view software development as a 'root-finding' procedure where a designer submits an initial 'guess', and iteratively tests for validity until a correct solution is converged upon.

Software is developed by human beings, and to err is human. In standard commercial software, errors are present, which are expensive. It is not possible to prevent these errors being introduced entirely, but can be located early. The ultimate goal of software testing must be customers' delight and their satisfaction, which can be achieved by finding out the maximum number of bugs before releasing the software. In order to achieve the goals of software testing a range objective are set including following:

- Identifying areas of weaknesses;
- Execute a program with the intent of finding an error in the weak areas identified;
- Generate a good test case that has a high probability of finding an as yet undiscovered error;
- Perform a successful test that uncovers an as yet undiscovered error;
- Provide confidence in the system;
- Establish the degree of quality;
- Establish the extent that the requirements have been met; and
- Provide an understanding of the overall system, and to prove that it is both usable and operable.

Good testing should be able to uncover different classes of errors in a minimum amount of time and effort. Software testing demonstrates that the software appears to be working as stated in the specifications. During testing, the data collected can also provide an indication of the software's reliability and quality. Software testing also identifies important defects, flaws, or errors in the application code that must be fixed. The testing team can only measure quality. They can't improve it. Although it is very true that designing tests before coding begins may improve quality because the coders can then use that information while thinking about their designs and during coding and debugging.

### 1.2.4 Challenges and Issues

Software testing presents very interesting anomaly for developers and testers. A major challenge is 'how to ensure exhaustive testing without actually doing this impossible task' for uncovering all the errors. However, this is unrealizable in reality but in order to maximize effect of testing, it is necessary to determine a good starting set of test cases that

- Eliminate redundant testing;
- Provide adequate test coverage;
- Allow more effective testing; and
- Make the most of limited testing resources.

Further, there are many problems in the process of software testing. Such problem for which no clear-cut solution/strategy exists is termed as issues. Some of the relative issues are described in the following section.

**Measurement of Testing Progress:** There is no clear strategy and foolproof technique to measure the progress in testing process, as the target of uncovering errors is not visible. However, only possible approach may be a variety of standard software metrics to determine whether the software under test has been tested enough.

**Voluminous Test-data:** Huge software with rapidly growing complexity leads to heavy test-data. Minimizing the size to manageable limit is needed to assure timely and effective testing. No clear strategy and technique, apart from sampling, exits.

**Late Deployment:** Testing process is generally considered as a part of coding activity and deployed late in software development life cycle. This is a mindset of most of the software developers. However, it may be deployed as early as possible in appropriate ways.

**Testability of Requirements:** A sizeable portion of software requirements may exist that may not be testable at all or less testable. Such portion may require creative strategy and techniques.

Apart from the challenges and issues discussed in foregoing section, there are repeatedly several problems with which testing process is comforted. Certainly, a little well-planned effort may help to solve these problems and enhance the quality, effectiveness, and efficiency of testing process. Some of the common problems are demonstrated in brief as follows:

**Misunderstanding the Role of Testing:** The major objective of software testing is to execute the test cases with the aim of finding the unknown errors/bugs in the given software application. Furthermore, it is important to have an understanding of the relative criticality of defects when planning tests, reporting status, and recommending actions.

**Poor Planning of the Testing Effort:** The objective of test plan is to find and report as many bugs as possible to improve the integrity of the program. Test plans often overemphasize testing functionality at the expense of potential interactions leading to incomplete configuration and inadequate load and stress testing. It is also risky decision to neglect the test documentation and/or installation procedures.

**Using the Wrong Personnel as Testers:** Testing should not be conducted by the programmer involved in developing the software under test, as the test result may be biased. A test group should include domain experts, and need not be limited to people who can program. A test team that lacks diversity will not be as effective.

**Poor Testing Methodology:** It is very important and quite difficult to choose the appropriate testing methodology and technique for the software under test. The testing method opted must verify that product does what it is supposed to, while not doing what it should not.

**Incomplete Testing:** No doubt, 100 percent testing can never be achieved. One of the major challenges to testers is that exhaustive testing is impossible. Several tests may be conducted to test the product, and hence several guidelines are available to achieve the optimal test. But, no doubt the scope of the tests may be maximized by undertaking appropriate measures. Some guidelines, for the purpose, are described as follows:

- Test should be conducted for every possible input to every variable.
- Test should be conducted for every possible combination of inputs to every combination of variables.
- Test should be conducted for every possible sequence through the program.

- Every hardware and software configuration, including configurations of servers should be tested.
- Every possible way the user might try to use the program should be tested.

**Misallocation of Resources:** Generally, testers misallocate resources because they fall for the company's process myths. This leads to a greater challenge to competent testing.

**Multiple Missions:** The test groups generally operate under multiple missions. The objectives of parallel and multiple missions often conflicts and rarely articulated.

**Unskilled Programmers:** Test groups often lack programming skills, and hence a vision of appropriate projects that would keep programming testers challenged.

## 1.3 EFFECTIVE SOFTWARE TESTING

Software testing activities play a critical role in the production of dependable systems, and account for a significant amount of resources including time, money, and personnel. Testing can only give a measure of confidence that a given software system is acceptable for its intended purpose. This level of confidence must be balanced against the role the system is intended for and against the risk of the system failing in operation, before the decision to release or to accept software can be made.

It is difficult to perform sufficient testing to be completely certain a given system is defect free. When this problem is combined with the fact that testing resources are finite and in short supply, then adequate testing becomes problematic. If testing effort can be more precisely focused on the places in a software system where faults are likely to occur, then available resources will be used more effectively and efficiently, resulting in more reliable systems produced at low cost.

The key to effective testing is making the process as efficient as possible. Testers must make use of their knowledge of testing techniques and experience to guide in devising effective testing. However, effective tests mean to uncover errors that are expected to, in right context. Entire testing process may be effectively conducted by going through a sequence of following phases as depicted in figure 1.3.

**Test Preparation (TEP):** The first and very basic approach towards testing is to set up areas to test, the test cases, test programs, and test environment. Amongst these, setting up the test environment can give a lot of trouble and delay, and hence an extra care should be taken on specifying all aspects of the test environment.

**Pre-Test (PRT):** This is the phase where test is run after the software under test is installed in the test lab with the help of few test cases. Pre-test is conducted with the objective whether the software under test is ready for testing at all, or totally unreliable or incompletely installed.

**Test (TES):** After implementing successful pre-test, the main test is carried out, which consists of all the pre-planned test cases. Test cases are run, and the failures are recorded and repaired as a result of testing observations or faults.

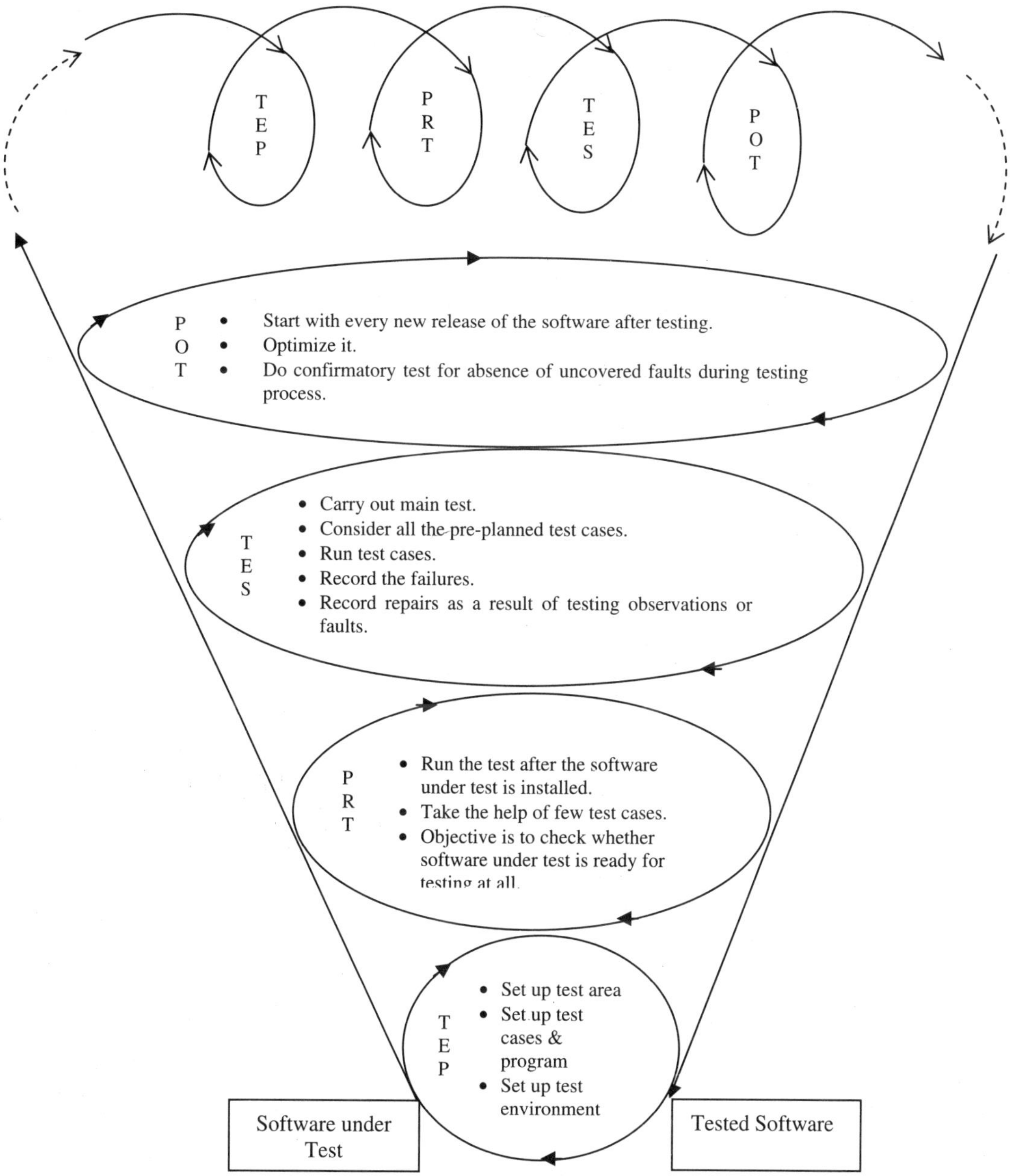


**Figure 1.3: Effective Software Testing Life Cycle**

**Post-Test(POT):** The major question arises what to do after testing the software, and when to optimize it. This phase starts with every new release of the software after testing. This is the phase where optimization should occur, and also may be at least taken as confirmatory test for absence of uncovered faults during testing process.

## ES OF TESTING

are testing is any activity aimed at evaluating an attribute or capability of a program or system and determining that it meets its required results. The difficulty in software testing stems from the complexity of software. A program with a moderate complexity cannot be completely tested. Software testing is a critical part of the software process. It comprises about 40 percent of the effort spent on the software process. There is a plethora of testing types that are classified by purposes in different life cycle phases. Few of the testing types are discussed in the following section:

**Black Box Testing:** The black-box approach is a testing technique in which test data are derived from the specified functional requirements without regard to the final program structure. Black box testing is not based on any knowledge of internal design or code. Tests are based on requirements and functionality. The tester treats the software under test as a black box, that is, only the inputs, outputs and specification are visible, and the functionality is determined by observing the outputs to corresponding inputs. In testing, various inputs are exercised and the outputs are compared against specification to validate the correctness.

**White Box Testing:** White box testing is based on knowledge of the internal logic of an application's code. In this type of the testing, software is viewed as a white-box, or glass-box, as the structure and flow of the software under test are visible to the tester. White box tests are based on coverage of code statements, branches, paths, and conditions.

**Unit Testing:** This testing is used to test particular functions or code modules. It is most 'micro' scale of testing. The developer carries out unit testing in order to check if the particular module or unit of code is working fine. Unit testing comes at the very basic level as it is carried out as and when the unit of the code is developed or a particular functionality is built. Since unit testing requires detailed knowledge of the internal program design and code, it is generally performed by the programmer and not by testers. Unit test is not always easily done, as it requires well-designed architecture of application with tight code.

**Integration Testing:** After all parts have been successfully tested, they are put together to form the complete software package, and then tested as a whole. This constitutes integration testing. The 'parts' can be code modules, individual applications, client and server applications on a network, etc. Integration testing depends mostly on black box testing and a little on white box testing. Integration testing is a systematic approach for constructing the program structure, while at the same time conducting tests to uncover errors associated with interfacing. This type of testing is especially relevant to client/server and distributed systems.

**Functional Testing:** Functional testing is a black-box type testing geared to functional requirements of an application. Functional testing is generally done by testers. This doesn't mean that the programmers shouldn't check that their code works before releasing it.

**Performance Testing:** Not all software systems have specifications on performance explicitly. But every system definitely has an implicit performance requirement. The software should not take infinite time or infinite resource to execute. Performance bugs sometimes are used to refer to those design problems in software that cause the system performance to degrade. The goal of performance testing can be performance bottleneck ide[illegible] performance comparison and evaluation. The typical method of doing performance [illegible] using a benchmark. Performance testing is done to ensure that processing has n[illegible] significantly degraded. It is not intended to test for optimizing the application.

**System Testing:** The system test is a more robust version of the external test, and can [illegible] known as an alpha test. System testing concentrates on testing the complete system with [illegible] variety of techniques and methods. The essential difference between 'system' and 'external function' testing is the test platform. System testing comes into picture after the unit and integration tests. In system testing, the platform must be as close to production use in the customers' environment, including factors such as hardware setup and database size and complexity. System testing is often combined with acceptance testing and/or integration testing.

**Regression Testing:** Regression testing is used to check that the changes made to a program have not introduced new faults in the system. Regression testing is an expensive but necessary activity performed on modified software. Regression testing has been receiving more attention as corporations focus on fixing the 'Year 2000 Bug'.

**Interoperability Testing:** Interoperability testing can be used to verify whether the renovated application can operate in its intended environment. This type of the testing requires a separate region or logical partition. It is very difficult to perform such type of testing, and requires much more effort. Interoperability testing is performed to test critical applications, and is combined with performance testing and implementation tests.

**Acceptance Testing:** Acceptance testing is the final testing based on specifications of the end-user or customer. An acceptance test is an exercise of a completed system by a group of end users to determine whether the system is ready for deployment. This testing is based on use by end-users/customers over some limited period of time. Here the system goes through more realistic testing that in the 'system test' phase, as the users have a better idea how the system will be used than the system testers.

**Volume Testing:** Volume testing examines the system's ability to perform correctly using large volumes of data with the aim of identifying defects that appear only under such conditions. It is sometimes termed as flood testing. Volume testing may involve rerunning appropriate test from earlier testing phases in addition to designing new test cases. One particular volume testing technique involves repeatedly running the same test while increasing the volume of the data involved at each iteration in order to identify the point at which the system fails to cope with the volume of data.

**Stress Testing:** Stress testing is often used interchangeably with 'load' and 'performance' testing. Some systems are designed to handle a specified load. Tests designed to ensure that the system can process its intended load is called stress testing. This testing also describes functional testing under heavy load, heavy repetition of certain actions or inputs, or large complex queries to a database system.

**Usability Testing:** Usability testing is used to test for 'user-friendliness'. This testing is subjective in nature, and hence will depend on the targeted end-user or customer. During rviews, surveys, video recording of user sessions, and other ility testing is an effective method for resolving internal debates y test establishes the ease of use and effectiveness of a product ctices. It also identifies usability problems with the product hose problems.

ation testing is often the most under tested area in testing. This med to ensure that all installed features and options function properly. ed to verify that all necessary components of the application are, indeed,

**very Testing:** Recovery testing tests how well a system recovers from crashes, hardware failures, or other catastrophic problems. Recovery testing is a system test that focuses the software to fall in variety of ways and verifies that recovery is properly performed. If it is automatic recovery then re-initialization, check pointing mechanisms, data recovery and restart should be evaluated for correctness. If recovery requires human intervention, the Mean-Time-To-Repair (MTTR) is evaluated to determine whether it is within acceptable limits.

**Security Testing:** Software quality, reliability and security are tightly coupled. Flaws in software can be exploited by intruders to open security holes. With the development of the Internet, software security problems are becoming even more severe. Many critical software applications and services have integrated security measures against malicious attacks. The purpose of security testing of these systems include identifying and removing software flaws that may potentially lead to security violations, and validating the effectiveness of security measures. Security testing is carried out in order to find out how well the system can protect itself from unauthorized access, hacking , cracking, any code damage etc., which deals with the code of the application. This type of testing needs sophisticated testing techniques, which itself is an area of research.

**Object-Oriented Testing:** Object-oriented systems are relatively difficult to test. The main problem with testing object oriented software is testing objects which are being reused. Testing object classes corresponds to unit testing. Since object-oriented programs emphasize encapsulation of internal object state and interaction between objects through method calls, it is natural to base a testing strategy on scenarios of object usage.

**User Acceptance Testing:** User Acceptance testing occurs just before the software is released to the customer. User acceptance testing determines if software is satisfactory to an end-user or customer. The end-users along with the developers perform the User Acceptance Testing with a certain set of test cases and typical scenarios.

**Alpha Testing:** Once the application has undergone system testing, it may be put through 'actual' use within the development organization by a customer. The software is in a natural setting with the developer recording errors and usage problems. This is called alpha or acceptance testing. Alpha testing is performed whenever an application development is nearing completion. As a result of this testing, minor design changes may still be made. This type of the testing is typically done by end-users or others, not by programmers or testers.

The alpha testing process continues until the system developers and client agree that the delivered system is acceptable. Alpha tests are conducted in a controlled environment.

**Beta Testing:** This test is conducted at one or more customer sites by the end user(s) of the software. The beta test is a 'live operational test' of the software in an environment that cannot be controlled by the developer. This testing is performed when development and testing are essentially completed and finally bugs and problems need to be found before final release. On the basis of feedback from the users, the developers determine whether any changes are necessary, before releasing the software product to the entire customer base.

**Mutation Testing:** Mutation testing is a kind of testing in which, the application is tested for the code that was modified after fixing a particular bug/defect. It also helps in finding out which code and which strategy of coding can help in developing the functionality effectively. Mutation testing is the process of re-writing source code in order to flush out ambiguities that may exist in the code. These ambiguities can cause failures in software if not detected and fixed by the developer. Because these faults are often very subtle, they can easily pass conventional testing and debugging only to be passed on to the customer. The process of Mutation testing is based upon seeding the implementation with a fault. This is possible by applying a *mutation operator*, and determining whether testing identifies this fault. A mutated program is called a *mutant* and if a test case distinguishes between the mutant and the original program it is said to *kill* the mutant. Mutation testing may be used to judge the effectiveness of a test set: the test set should kill all the mutants. Mutation analysis is a fault-based technique with the aim of measuring the mutant of test data. It involves constructing a set of mutant programs, each of which is a copy of initial program under test that differs by exactly one mutation. A mutation may be defined as the single, syntactically correct change that is made by applying a predefined transformation rule, called a mutation operator, to the original program.

**Reliability Testing:** Reliability testing involves tests to ensure the robustness and reliability of the application under test, under typical usage. It typically includes integrity tests, which focuses on verifying the application's robustness and compliance to language, syntax, and resource usage. Reliability testing also includes structural tests, which focuses on verifying application's adherence to its design and information.

**Regression Testing:** A regression test, also known as validation testing in literature, provides a consistent, repeatable validation of each change to an application under development or being modified. A concern that should be kept in mind during testing is that software will be modified during its lifetime. For example, new functionality may be added to the software, or a module may be modified to improve its response time. The changes may introduce errors into the software that was previously correct. Regression testing involves retesting of an application or system that has been modified to insure that no previously working components, functions, or features fail as a result of the repairs. Regression testing is conducted in parallel with other tests. Regression testing ensures whether the newly modified code complies with its specified requirements, as well as unmodified code has not been affected by the change. Regression testing doesn't test that a specific defect has been fixed, rather, it tests that the rest of the application up to the point or repair was not adversely affected by the fix.

**Back-Out Testing:** This type of the testing is used to verify that a revised application can be taken out of production and the previous version can be reinstalled. Its major objective is to prove that a fail-safe mechanism exists and can be invoked, if necessary. This test is typically run back-to-back with an implementation test. Back-out testing is not usually combined with any other test type.

**Penetration Testing:** It is a kind of search for vulnerabilities from software or a computer system, and a system is said to be penetrated whenever vulnerability is found and exploited. Generally, a specialized team called tiger team does penetration testing. During penetration testing, the tester can scan for vulnerabilities which have been found earlier from similar systems. Penetration testing is based heavily on the experience of the tester. Penetration testing can also be based on the systematic creation of malicious input data fed to a tested software component. The search for known vulnerabilities can be manual. Several security scanners are also available to be used as a tool for penetrating the software system. Security scanners are based on databases containing known vulnerabilities.

**String Testing:** String testing can be used for small or logical grouping of programs that are tightly interrelated, which makes the program difficult to be tested individually. Usually, this type of testing consists of executing an on-line transaction or batch job to verify that it functions correctly. Generally, string testing is combined with unit testing, and sometimes with system testing.

## 1.5 PRINCIPLES OF SOFTWARE TESTING

Bugs persist in contemporary software and cause inconvenience and occasional loss of information. Quality is evaluated by software testing to reduce faults in released software. Traditionally, software testing attempts to ensure that the software meets the specifications rather than trying to find vulnerabilities in software. Software testing is an extremely creative and intellectually challenging task. Software testing can be made effective and efficient by following the established principles in this regard. A selection of such principles is briefly described as follows: the principles given below:

**Testing must be done by an independent party:** Actual software testing is an independent activity, and may not be performed by the person or team that developed the software. They tend to defend the correctness of the program and also overlook own errors, as a natural human behavior. The programmer who wrote the application should have a reduced role in the testing because they may not be able to take an unbiased look at the results of their labors.

**Assign best personnel for testing:** Software testing requires high level creativity and responsibility. Therefore, only the best personnel in the area must be assigned to design the test cases. They must have considerable degree of expertise in implementing and analyzing the test cases, test data and test results.

**Test should be planned with the objective of getting maximum number of errors:** Good testing should be able to uncover maximum number of errors in a minimum amount of time and with a minimum amount of effort. The objective of testing should be to show that the software defects are present. It should not show the absence of errors. Therefore, test process should be planned accordingly.

**Invalid and unexpected input conditions as well as valid conditions should be tested:** All possible invalid and unexpected input conditions should be checked during testing with the objective of getting correct message. The program should also be tested to generate correct result when a valid input is considered.

**Software under test should not be modified during testing:** Modification of the module under test may not give the appropriate testing result. Therefore, during execution of designed test cases, the module under test should not be modified.

**Prepare test report including test cases and test results:** To summarize the results of the designated testing activities and to provide evaluations based on these results, a test report will be prepared at last. This report will provide a brief overview of the tested software and summarize the testing effort along with the release version, number of test phases, number of rounds of tests executed within each test phase, and the overall duration of the entire testing effort.

**Identify and include expected results if possible:** Expected result for the requirement specification is a necessary part of test documentation. Even though, it is very impractical to identify the desired output. It is highly desirable so that results may be verified and categories as 'correct or wrong' leading to the knowledge of errors.

**Test should be planned in advance and should be updated later:** Software test plan will map the tests to the requirements and define the entry and exit criteria for each phase of testing. It will define the test system needed for testing, and the work product for each phase of testing. The test plan should be written before the module under consideration is begun, and should be revised to add to and refine the tests.

**Choose an appropriate set of test methods:** Different parts of a system with different characteristics and different levels of importance will attract different testing methodologies. Therefore, appropriate test methods should be chosen for testing parts of a system.

## 1.6 TESTING AND DEBUGGING

In the last 50 years, it has been seen tremendous productivity increases in all areas of software development. In commercial software development organizations, increased complexity of products, shortened development cycles, and higher customer expectations of quality have placed a major responsibility on the areas of software debugging, testing, and verification. Irrespective of the specific development model followed, certain basic activities are required for a successful development and execution of a software product. Figure 1.6 (a) illustrates the specific roles played by debugging, testing, and verification during software development life cycle.

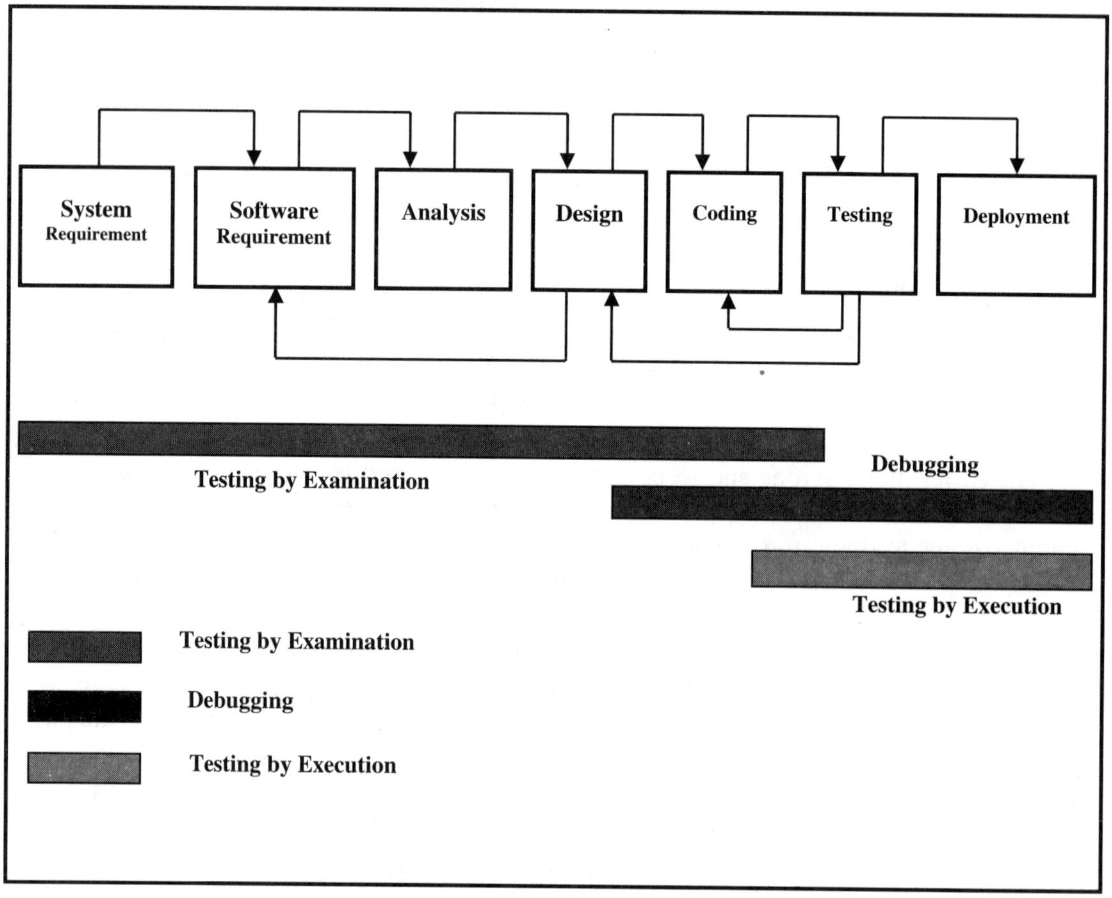


**Figure 1.6 (a): Testing and Debugging integrated with Software Development Process**

Major purpose of software testing is to increase the software engineer's confidence in the proper functioning of the software. It is the process of dynamic verification of program's behavior on finite set of test cases selected from infinite execution domain. Since the objective of testing is to list out errors, a successful test is one that uncovers maximum number of undiscovered errors. The tester aim is to design tests that systematically uncover different classes of errors with a minimum amount of time and effort. Software testing is a process that can be systematically planned and specified. Test design can be conducted, a strategy can be defined, and results can be evaluated against prescribed expectations. Software testing provides technique to reveal bugs that would be too costly to find with other verification and validation techniques. A bug or error in software product is an exception that can hinder the functionality of either the whole software or part of it. Test cases or scripts are run in order to find out any unexpected behavior of the software under test. If any such unexpected behavior or exception occurs, it is called a bug. After the bug or error is detected in software under test, it needs to be communicated to the developer or test leader. The bug reported will be verified whether it is valid or not in order to get it fixed.

Testing is the process of executing software with the intent of detecting the presence of faults. Testing provides a measure for quality. Testing is clearly a necessary area for software validation. Typically, prior to coding the program, design reviews and code inspections are done as part of the static testing effort. Once the code is written, various other static analysis methods based on source code can be applied. The testing done at later stages is 'black box' testing, which is based on external specifications, and hence does not involve the understanding of the detailed code implementations. Typically, system testing targets key aspects of the product, such as recovery, security, stress, performance, hardware configurations, software configurations, etc. Testing during production and deployment typically involves some level of customer-acceptance criteria.

Debugging refers to undertaking measures to make programs free from bugs. It is an integral component of software testing process. Debugging occurs as a consequence of successful testing and revealing the bugs from the software under test. When a test case uncovers an error, debugging is the process that results in the removal of the bugs. Debugging is not testing, but it always occurs as a consequence of testing. The debugging process begins with the execution of a test case. The debugging process attempts to match symptom with cause, thereby leading to error correction. The purpose of debugging is to locate and fix the offending code responsible for a symptom violating a known specification.

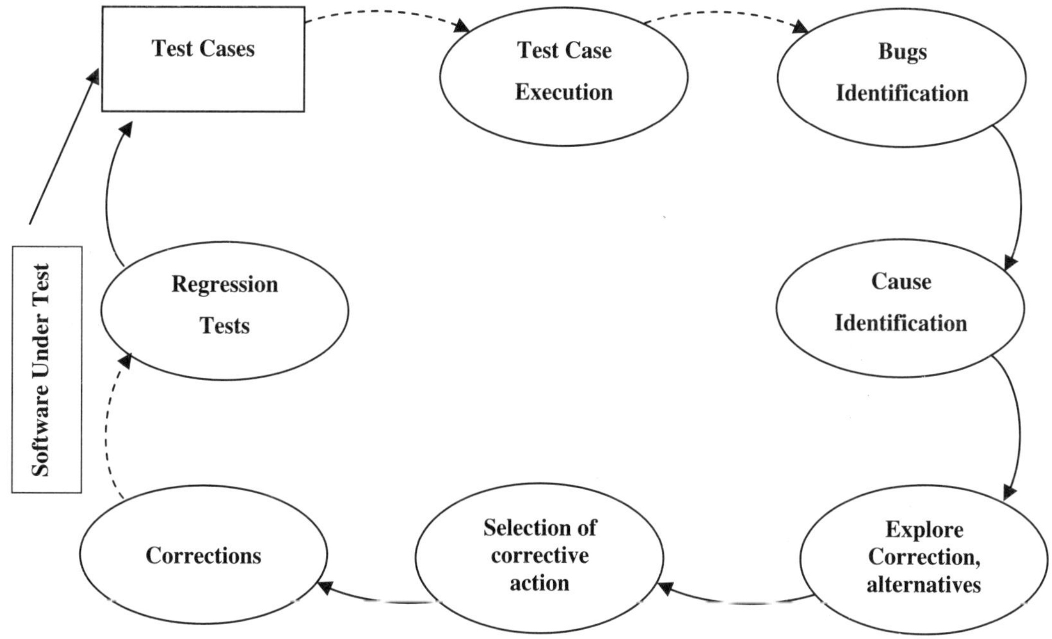


**Figure 1.6 (b): Debugging Life Cycle**

Figure 1.6 (c) depicts the contribution of software testing and debugging in fault detection.

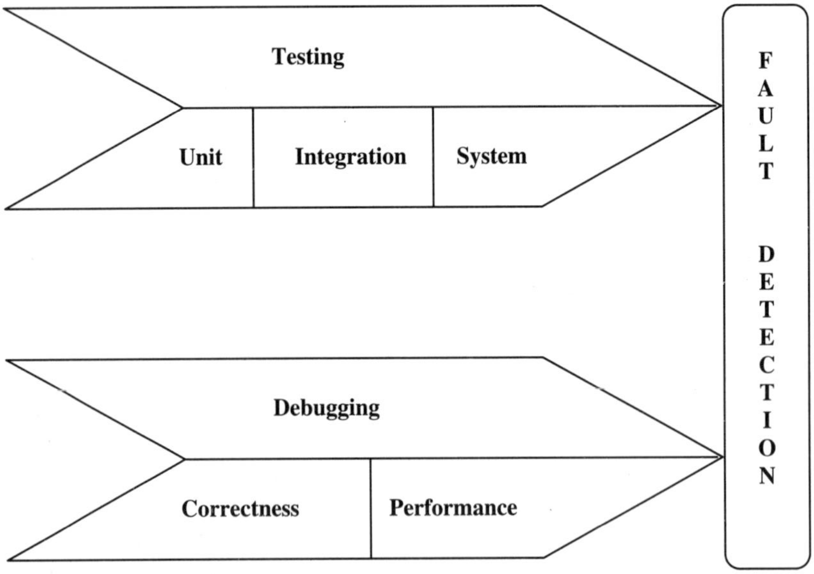


**Figure 1.6(c): Role of Testing and Debugging in Fault Detection**

Software bugs can be described by three perspectives including *Error*, *Defect*, and *Failure*. Errors are the mistakes made by a developer, and are mostly located in people's head. Software errors may lead to one or more defects, which are located in text of program. Execution of defected code may lead to failures.

**Error:** The system is in such a state that further processing by the system will lead to a failure.

**Defect:** The mechanical or algorithmic cause of an error.

**Failure:** Any deviation of the observed behavior from the specified behavior.

Three perspectives of bugs are shown in figure 1.6 (d). Table 1.0 shows the examples of different kinds of bugs.

**Table 1.0: Examples of Different Types of Bugs**

| Interface specification Bugs | Algorithmic Bugs | Mechanical Bugs |
|---|---|---|
| • Mismatch between what the client needs and what the server offers<br>• Mismatch between requirements and implementation | • Missing initialization<br>• Branching errors (too soon, too late)<br>• Missing test for nil | • Documentation does not match actual conditions or operating procedures |

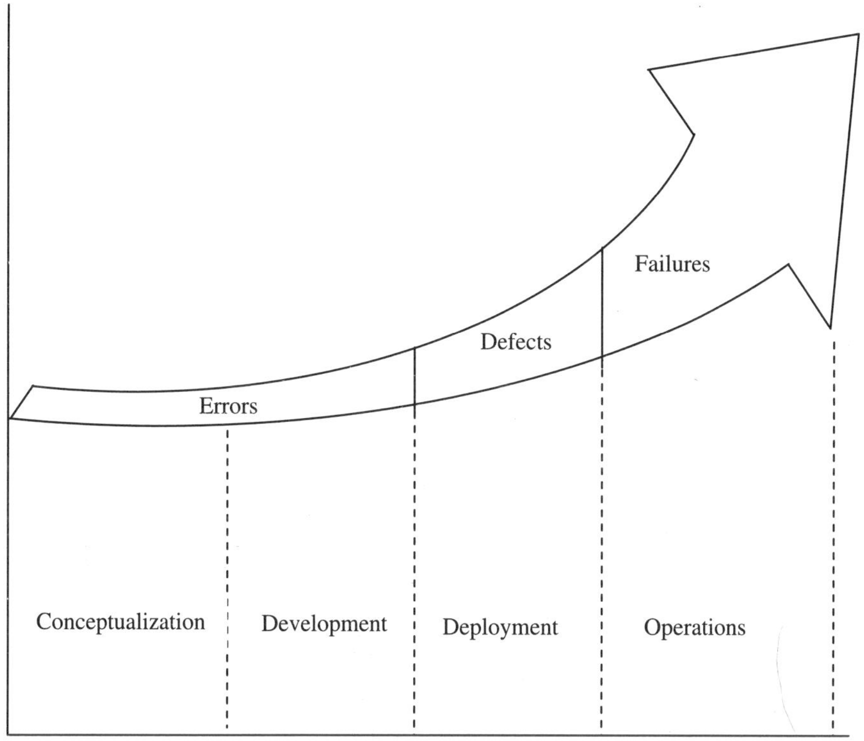


**Figure 1.6(d): Software Bugs-Different Perspectives**

Following characteristics of the bugs has been identified for the software under test:

(i) The bug revealed may temporarily disappear when another bug is revealed and corrected.

(ii) The revealed bugs may be caused by nonerrors.

(iii) The bugs may be caused because of the human errors, which is very difficult to trace out.

(iv) The bugs revealed may be because of the timing problem, rather than processing problem.

(v) Reproduction of input conditions accurately may be difficult.

A software programmer, evaluating the results of a test, is often confronted with a symptomatic indication of a software problem. Debugging is the process of finding and correcting the cause of software bugs. Several approaches have been discussed in literature for debugging software under test, including the following three:

*Brute Force Approach:* Brute force approach is the most common approach used when all else fails. This is the least effective method for isolating the cause of software errors. Lots of thoughts, much more effort and time are required while implementing this approach.

*Backtracking Approach:* This approach is usually applied and successful for comparatively small programs. As the symptoms have been uncovered, the source code is traced backward manually until the cause is found.

*Cause Elimination Approach:* This approach introduces the concept of binary partitioning using induction and deduction.

Debugging typically happens during the following three phases of software development life cycle, and the level of granularity of the analysis required for locating the defect differs in these three.

***Coding Phase:*** In coding phase, when the programmer translates the design into an executable code, the errors made by the programmer in writing the code can lead to defects that need to be quickly detected and fixed before the code goes to the next stages of development. Most often, the developer also performs unit testing to expose any defects at the module or component level.

***Testing Phase:*** A certain amount of debugging of the test execution is necessary to conclude that the program under test is the cause of the unexpected behavior and not the result of a bad test case due to incorrect specification, inappropriate data, or changes in functional specification between different versions of the system. Once the defect is confirmed, debugging of the program follows and the misbehaving component and the required fix are determined.

***Deployment Phase:*** Some undesirable aspects of software behavior, such as inadequate performance under a severe workload or unsatisfactory recovery from a failure, get exposed at deployment phase and hence the offending code needs to be found and fixed before large-scale deployment.

## 1.7 CONCLUSION

Software testing is a process of verifying and validating software application or program to achieve the business and technical requirements. One of the major objectives of the testing is to generate a good test case that has a high probability of finding an as yet undiscovered different classes of errors in a minimum amount of time and with a minimum amount of effort. Software testing demonstrates that the software appears to be working as stated in the specifications. One of the purposes of software testing is to exercise and observe the product behavior through dynamic verification. A comprehensive software testing, inspection, analysis and other verification tasks performed at each stage of software development life cycle will decide whether the software is validated, and the test used for this purpose is named clean tests, or positive tests.

To test software quality, its related factors, which make quality visible, should be tested directly. Ample number of test cases may be written to test a small piece of software, but the only problem is to select the minimum set of test case sufficiently effective for revealing potential faults in a program. It is evident from the literature survey that the evaluation criterion for test strategies may be used to measure the effectiveness of generated test cases.

Neglecting test documentation and installation procedures may be risky. A test team is carrying out software testing, and therefore, a test team that lacks diversity will not be as effective.

An appropriate test method must be able to verify whether the product does what it is supposed to. The programmer involved in developing the software should have a reduced role in the testing because they may not be able to take an unbiased look at the results of their labors. Software testing has been classified by the purposes in different life cycle phases. Black box tests are based on requirements and functionality, while white box tests are based on coverage of code statements, branches, paths, and conditions. Unit testing is carried out as and when the unit of the code is developed or a particular functionality is built. Integration testing depends mostly on black box testing and a little on white box testing, and is a systematic approach for constructing the program structure, while at the same time conducting tests to uncover errors associated with interfacing.

Functional testing is a black-box type testing geared for functional requirements of an application, with the goal of identifying bottlenecks, performing comparison and evaluation. System testing comes into picture after the unit and integration tests. Regression testing is used to check that the changes made to a program have not introduced new faults in the system. An acceptance test is an exercise of a completed system by a group of end users to determine whether the system is ready for deployment. During usability testing, user interviews, surveys, video recording of user sessions, and other techniques can be used.

Recovery testing is a system test that focuses the software to fall in a variety of ways and verifies that recovery is properly performed. Security testing is carried out in order to find out how well the system can protect itself from unauthorized access, hacking – cracking, any code damage etc. which deals with the code of application. User acceptance testing determines if software is satisfactory to an end-user or customer. Alpha testing is performed whenever an application development is nearing completion, and as a result, minor design changes may still be made. The beta test is a 'live' application of the software in an environment that cannot be controlled by the developer, and is performed when development and testing are essentially completed and final bugs and problems need to be found before final release.

## KEY TERMS

Testing, Software testing, Test plan, Test process, Testing objectives, Good testing, Verification, Validation, Defect prevention, Quality Improvement, Reliability Estimation, Mutation analysis, poor planning, testing efforts, Test group, Test team, Testing methodology, Incomplete test, Effective testing, Test preparation, Pre-test, main test, after test, black box testing, White box testing, Unit testing, Integration testing, Functional testing, Performance testing, System testing, Regression testing, Acceptance testing, Stress testing, Usability testing, installation testing, Recovery testing, Security Testing, Object oriented testing, User acceptance testing, Alpha testing, Beta testing, Mutation testing, Penetration testing, Software Testing principles.

## REVIEW QUESTIONS

### OBJECTIVE TYPE QUESTIONS

1. Software testing is the process of:
   (a) Verifying software application
   (b) Validating software application
   (c) Verifying and validation software application
   (d) None of the above
2. Good test should uncover errors in:
   (a) Minimum amount of time
   (b) Minimum amount of effort
   (c) Minimum amount of time and effort
   (d) None of the above
3. Testing team can:
   (a) Measure quality
   (b) Can improve quality
   (c) Cannot measure quality
   (d) Cannot improve quality
4. Software testing comes under
   (a) Static verification
   (b) Dynamic verifications
   (c) Both static and dynamic verifications
   (d) None of the above
5. A variance between the expected result and actual result is called:
   (a) Defect
   (b) Fault
   (c) Bug
   (d) Failure
6. Software quality can be:
   (a) Tested directly
   (b) Cannot be tested directly
   (c) Predicted on the basis of software quality factors
   (d) None of the above

7. Software testing starts with
   (a) Requirements
   (b) Code
   (c) Design
   (d) None of the above
8. Software testing is:
   (a) Static activity
   (b) Dynamic activity
   (c) Both static and dynamic activity
   (d) None of the above
9. System testing comes after:
   (a) Integration testing
   (b) Unit testing
   (c) Integration and Unit testing
   (d) None of the above
10. ______________ is final testing based on specification of the end user or customer:
   (a) Regression testing
   (b) Acceptance testing
   (c) Functional testing
   (d) Performance testing
11. Stress testing is often used interchangeably with:
   (a) Load testing
   (b) Performance testing
   (c) Load and performance testing both
   (d) None of the above
12. Testing object class corresponds to:
   (a) Unit testing
   (b) System testing
   (c) Integration testing
   (d) Acceptance testing
13. ____testing is performed whenever an application development is nearing completion.
   (a) Acceptance testing
   (b) Alpha testing

(c) Beta testing
(d) System testing

14. Good testing should be able to:
(a) Uncover maximum number of errors
(b) Minimum number of errors
(c) Reduce cost and efforts
(d) None of the above

## SHORT ANSWER TYPE QUESTIONS

1. Why usually developers spend 40 percent of total project time on testing?
2. Testing is often considered as an expensive and uncontrollable process. Why? Justify your answer.
3. What is software testing? Describe briefly.
4. What do you mean by test planning?
5. Define software test process.
6. Write down the characteristics of good testing.
7. Write down the responsibilities of software testing tools.
8. Write down the need and importance of software testing.
9. Identify the component of software testing.
10. Describe briefly the general purpose of software testing.
11. Distinguish between software verification and validation.
12. What do you mean by specifications?
13. Distinguish between static verification and dynamic verification.
14. How software inspection is related with static verification?
15. How to conclude that the software is valid?
16. What do you mean by dirty test or negative test?
17. What do you mean by quality end software?
18. Identify the importance and significance of bugs in software system.
19. List out the objectives of debugging.
20. Find out the commonly accepted set of software quality factors.
21. Distinguish between faults and failures.
22. What do you mean by software test strategy?

23. What is mutation analysis? How it assists in quality of test case assessment?
24. Identify at least five set of characteristics of software testing not included in this chapter.
25. Write down the importance of test group during software testing.
26. What do you mean by effective testing?
27. How to set up an appropriate test environment during testing?
28. Distinguish between white box and black box testing.
29. Unit testing is performed by the programmer, not by the testers. Why?
30. What is integration testing?
31. Who performs functional testing?
32. Identify the goals of performance testing.
33. Regression testing is an experience but necessary activity performed on modified software. Justify.
34. Write down the purpose of security testing.
35. How object oriented testing deviates from testing traditional software testing?
36. Distinguish between alpha and beta testing.

## DESCRIPTIVE QUESTIONS

1. Establish a correlation between software testing and software reliability estimation.
2. List out the challenges with regard to software testing. Propose the solutions to meet out the challenges discussed.
3. 'A finite number of tests cannot validate that the software works for all situations'. Justify the statement.
4. Write down the steps for defect prevention. How it differs from defect detection?
5. How software testing is related with software quality assurance? How much software testing contributes in improving software quality?
6. 'Debugging is a narrow view of software testing, which can be performed heavily to find out design defects'. Justify the statement.
7. Suggest a guideline to select a minimum set of test cases sufficiently effective for revealing potential faults in a program.
8. How poor planning of testing effort affects overall software testing. Suggests guidelines to improve the planning of testing effort?
9. Write down the steps while assigning the testing task to the personnel.
10. Prepare a guideline to select the appropriate methodology for testing software.

## REFERENCES

1. T. Ball, R. Majumdar, T. Millstein, S.K. Rajamani, Automatic Predicate Abstraction of C Programs, Proc. PLDI 2001, f2SIGPLAN Notices, Vol. 36, No. 5, pp. 203-213.
2. G. Brat, K. Havelund, S. Park, W. Visser, Java PathFinder A 2nd generation of a Java model checker, Proc. Workshop on Advances in Verification, Chicago, Ill., July 2000.
3. E.M. Clarke, O. Grumberg, and D. Peled, Model Checking, MIT Press, January 2000.
4. J.C. Corbett, M.B. Dwyer, et al., Bandera: Extracting Finite-State Models From Java Source Code, Proc. 22nd Int. Conf. on Software Engineering, June 2000, pp. 439-448.
5. K. Havelund, and T. Pressburger, Model Checking Java Programs Using Java PathFinder, Int. Journal on Software Tools for Technology Transfer, Vol. 2, No. 4, April 2000, pp. 366-381.
6. T.A. Henzinger, R. Jhala, et al., Software Verification with Blast, Proc. 10th SPIN Workshop on Model Checking Software, LNCS 2648, Springer-Verlag, 2003.
7. G.J. Holzmann, The Model Checker SPIN, IEEE Trans. on Software Engineering, Vol. 23, No. 5, May 1997, pp. 279-295.
8. G.J. Holzmann, and M.H. Smith, An Automated Verification Method For Distributed Systems Software Based On Model Extraction, IEEE Trans. on Software Engineering, Vol. 28, No. 4, April 2002, pp. 364-377.
9. G.J. Holzmann, and M.H. Smith, FeaVer 1.0 User Guide, Bell Labs, Dec. 2002, 64 pgs. Online document http://cm.bell-labs.com/cm/cs/what/modex/.
10. G.J. Holzmann, The SPIN Model Checker: Primer and Reference Manual, Addison-Wesley, ISBN 0-32122-862-6, August 2003.
11. M.L. Minsky, Computation: Finite and Infinite Machines, Prentice Hall, Englewood Cliffs, N.J., 1967.
12. Pnueli, The Temporal Logic Of Programs. Proc. 18th IEEE Symp. on Foundations of Computer Science, 1977, Providence, R.I., pp. 46-57.
13. Strachey, An Impossible Program, Computer Journal, Vol. 7 No. 4, Jan. 1965, p. 313.
14. A.M. Turing, On Computable Numbers, With An Application To The Entscheidungs Problem, Proc. London Mathematical Soc., Ser. 2-42, 1936, pp. 230-265.
15. Boris Beizer, Black Box Testing, Techniques For Functional Testing Of Software And Systems, John Wiley & Sons, 1995. ISBN 0-471-12094-4.
16. Boris Beizer, Software System Testing and Quality Assurance, International Thomson Computer Press, 1996. ISBN 1-85032-821-8.

17. Boris Beizer, Software Testing Techniques, Second Edition, Van Nostrand Reinhold, 1990. ISBN 0-442-20672-0.
18. Richard Bender, Writing Testable Requirements, Version 1.0, Bender & Associates, Inc., Larkspur, CA 94777, 1996.
19. Frederick P. Brooks, Jr., The Mythical Man-Month, Essays on Software Engineering, Addison- Wesley Longman, Anniversary Edition, 1995. ISBN 0-201-83595-9.
20. Silvana Castano, et.al., Database Security, ACM Press, Addison-Wesley Publishing Company, 1995. ISBN 0-201-59375-0.
21. Computerized Data Systems for Nonclinical Safety Assessment, Current Concepts and Quality Assurance, Drug Information Association, Maple Glen, PA, September 1988.
22. M. S. Deutsch, Software Verification and Validation, Realistic Project Approaches, Prentice Hall, 1982.
23. Robert H. Dunn and Richard S. Ullman, TQM for Computer Software, Second Edition, McGraw- Hill, Inc., 1994. ISBN 0-07-018314-7.
24. Elfriede Dustin, Jeff Rashka, and John Paul, Automated Software Testing – Introduction, Management and Performance, Addison Wesley Longman, Inc., 1999.
25. Robert G. Ebenau and Susan H. Strauss, Software Inspection Process, McGraw-Hill, 1994. ISBN 0-07-062166-7.
26. Richard E. Fairley, Software Engineering Concepts, McGraw-Hill Publishing Company, 1985. ISBN 0-07-019902-7.
27. Michael A. Friedman and Jeffrey M. Voas, Software Assessment - Reliability, Safety, Testability, Wiley-Interscience, John Wiley & Sons Inc., 1995.
28. Tom Gilb, Dorothy Graham, Software Inspection, Addison-Wesley Publishing Company, 1993. ISBN 0-201-63181-4.
29. Robert B. Grady, Practical Software Metrics for Project Management and Process Improvement, PTR Prentice-Hall Inc., 1992. ISBN 0-13-720384-5.
30. Les Hatton, Safer C: Developing Software for High-integrity and Safety-critical Systems, McGraw-Hill Book Company, 1994. ISBN 0-07-707640-0.
31. Janis V. Halvorsen, A Software Requirements Specification Document Model for the Medical Device Industry, Proceedings IEEE SOUTHEASTCON '93, Banking on Technology, April 4th -7th, 1993, Charlotte, North Carolina.
32. Debra S. Herrmann, Software Safety and Reliability: Techniques, Approaches and Standards of Key Industrial Sectors, IEEE Computer Society, 1999. ISBN 0-7695-0299-7.
33. Bill Hetzel, The Complete Guide to Software Testing, Second Edition, A Wiley-QED Publication, John Wiley & Sons, Inc., 1988. ISBN 0-471-56567-9.

34. Watts S. Humphrey, A Discipline for Software Engineering. Addison-Wesley Longman, 1995. ISBN 0-201-54610-8.

35. Watts S. Humphrey, Managing the Software Process, Addison-Wesley Publishing Company, 1989. ISBN 0-201-18095-2.

36. Capers Jones, Software Quality, Analysis and Guidelines for Success, International Thomson Computer Press, 1997. ISBN 1-85032-867-6.

37. Better Software magazine. Free issue at Hhttp://www.zinio.com/offer?issn=1532-3579f&of=PF01&bd=1H

38. Ergo/Gero Hhttp://www.ergogero.com/FAQ/Part4/cfaqPart4.htmlH (information on color blindness)

39. Goldsmith, Robin F. (2002). Software Development Magazine, 4-part series, July-October. Information on the V-Model is at http://www.sdbestpractices.com/documents/s=8815/sdm0208e/

40. International Institute for Software Testing. Hhttp://www.testinginstitute.com/H

41. Marrik, Brian. "Classic Testing Mistakes" (1997) and "New Models for Test Development" (1999). http://www.testing.com/writings/writings.html

42. Parasoft Corporation. "Automated Error Prevention". http://www.parasoft.com/jsp/aep/aep.jsp?itemId=170

43. J.M. Juran, Frank M. Gryna, Quality Planning and Analysis, Third Edition, , McGraw-Hill, 1993. ISBN 0-07-033183-9.

44. Stephen H. Kan, Metrics and Models in Software Quality Engineering, Addison-Wesley Publishing Company, 1995. ISBN 0-201-63339-6.

45. Cem Kaner, Jack Falk, Hung Quoc Nguyen, Testing Computer Software, Second Edition, Vsn Nostrand Reinhold, 1993. ISBN 0-442-01361-2.

46. Craig Kaplan, Ralph Clark, Victor Tang, Secrets of Software Quality, 40 Innovations from IBM, McGraw-Hill, 1995. ISBN 0-07-911795-3.

47. Edward Kit, Software Testing in the Real World, Addison-Wesley Longman, 1995. ISBN 0-201- 87756-2.

48. Alan Kusinitz, "Software Validation", Current Issues in Medical Device Quality Systems, Association for the Advancement of Medical Instrumentation, 1997. ISBN 1-57020-075-0.

49. Nancy G. Leveson, Safeware, System Safety and Computers, Addison-Wesley Publishing Company, 1995. ISBN 0-201-11972-2.

50. Michael R. Lyu, Editor, Handbook of Software Reliability Engineering, IEEE Computer Society Press, McGraw-Hill, 1996. ISBN 0-07-039400-8.

51. Steven R. Mallory, Software Development and Quality Assurance for the Healthcare Manufacturing Industries, Interpharm Press,Inc., 1994. ISBN 0-935184-58-9.

52. Brian Marick, The Craft of Software Testing, Prentice Hall PTR, 1995. ISBN 0-13-177411-5. Steve McConnell, Rapid Development, Microsoft Press, 1996. ISBN 1-55615-900-5.

53. Glenford J. Myers, The Art of Software Testing, John Wiley & Sons, 1979. ISBN 0-471-04328-1.

54. Peter G. Neumann, Computer Related Risks, ACM Press/Addison-Wesley Publishing Co., 1995. ISBN 0-201-55805-X.

55. Daniel Olivier, Conducting Software Audits, Auditing Software for Conformance to FDA Requirements, Computer Application Specialists, San Diego, CA, 1994.

56. William Perry, Effective Methods for Software Testing, John Wiley & Sons, Inc. 1995. ISBN 0- 471-06097-6.

## USEFUL LINKS

http://www.softwaretechnews.com/stn3-3/final-frontier.html

http://www.bullseye.com/coverage.html

http://www.elet.polimi.it/conferences/fase06/

http://www.hicss.hawaii.edu/hicss_40/fstcfp.htm

http://jan.netcomp.monash.edu.au/itp/l05.html

http://mingo.info-science.uiowa.edu/soft-eng/

http://satc.gsfc.nasa.gov/assure/agb.txt

http://cs.wwc.edu/~aabyan/435/Construction.html

http://www.sqe.com/downloads/stareast.pdf

http://www.softtest.org/sigs/material/dimaggio5.htm

http://www.url.edu/en/ip/center/ lasalle/subjects/ing_50043_TM.pdf

http://www.berlin.ptb.de/8/83/831/swq/ansqs/swp_vale.html

http://www.satisfice.com/articles/et-article.pdf

http://www.engineeringtalk.com/news/abg/abg101.html

ftp.cwi.nl/CWIreports/AA/CS-R9343.pdf

http://misrc.umn.edu/workshops/ 2002/fall/Rothenberger_111502.pdf

http://www.csis.ul.ie/Modules/CS5703/BlackBox-Triangle.ppt

http://www.codecomments.com/archive386-2006-3-841539.html

www.testingeducation.org/a/fundchal.pdf

www.testingeducation.org/articles

www.mindfiresolutions.com/ download/Checklists=Organization.pdf
www.scantron.com/downloads/pdf/CAT-Paper.pdf
lcgapp.cern.ch/project/processV1.html
quality-assurance-software-testing.blogspot.com/ 2005/08/rapid-testing.html
www.veritest.com/services/qa.asp
www.ccs.neu.edu/home/lieber/ com3220/lectures/lec1-test.PDF
www.evolutif.co.uk/courses/cdfcoursebrochure.pdf

# Chapter 2

# Software Testability

*The true test of character is not how much we know how to do, but how we behave when we don't know what to do.*

**John W. Holt**

## OBJECTIVES

The aim of this chapter is:

- To understand and describe software testability in general.
- To appreciate the need, importance and significance of software testability.
- To conceptualize the software testability estimation framework.
- To identify the role of software testability for software testing.
- To understand the concepts of software testability improvements.
- To realize the facts related to various features of software testability metrics.
- To enumerate and realize efforts made in the development of software testability models.
- To identify and comprehend and identify the software testability facilitators.
- To understand the fish-bone model of software testability.
- To identify and list software quality factors.
- To list common and distinguishing features of software.
- To relate software characteristics with testability.
- To estimate efforts made in mapping software quality factors and software characteristic.
- To understand and describe software testability heuristics.
- To realize the facts related with design for testability.
- To visualize the selection of software testability factors.
- To perform software testability analysis, through the available software components.
- To estimate the level of testability of software.

## 2.1 INTRODUCTION

Software testing is one of the most crucial activities of its development life cycle. It is used to assess the compliance of a program to its intended specifications and to assess the reliability of the program to inputs, which were not intended to be part of the specifications. Any technique that improves a software design at an early stage of software development life cycle can have highly beneficial impact on the final testing cost and efficiency. Good test inputs are those that result in a high probability of discovering errors upon program execution.

Object orientation is now widely chosen by software industry, despite that the technology is not mature enough from the testing point of view. Testing of object-oriented software presents new challenges due to features such as inheritance, polymorphism, dynamic binding, and object state. Programs contain complex interactions among the sets of collaborating objects from different classes. These interactions are greatly complicated by object-oriented features such as polymorphism, which allows the binding of an object reference to objects of different classes. While this is a powerful mechanism for producing compact and extensible code, it creates several fault holes. Though a class diagram is often termed as ambiguous, incomplete, it may lead to several false interpretations. Consequently it may possibly lead to false implementations and useless tests. It appears as a good basis to detect and master the widespread implicit control dependencies, due to inheritance and dynamic binding to guide the testing task.

One of the important goals of software developers is to keep the maintenance effort under control. A major factor, determining the maintenance effort, is the software system and its degree of maintainability. Maintainability is a defined as the set of attributes that bear on the effort needed to make specified modifications. One of the sub characteristics of maintainability is testability, which is defined as a *set of attributes that bear on the effort needed for validating the modified software*. Software testability is one of important concepts in design and testing of software program and components. Building programs and components with good testability always simplifies test operations, reduces test cost, and increases software quality. One way to improve the maintainability of a software system is design for testability, which can address different aspects of software including size, complexity, system structure, built-in-test facilities, distribution, and non-determinism.

Design for testability is a strategy to align the development process so that testing is maximally effective under either a reliability-driven or resource limited regime. There are several set of program characteristics that lead to testable software including operability, observability, controllability and understandability. Software testability analysis may be useful to examine and estimate the quality of software testing using an empirical analysis approach. Testability is important for both ad-hoc developers and organizations with a high-level of process maturity; it reduces cost in a reliability-driven process and increases reliability in resource-limited processes. It refers to 'the inherent ability or extent of ease with which software undergoes through testing'.

Object oriented system presents some unique obstacles to testability as well as sharing many with conventional implementations. The existence of the client-supplier relationships in the system causes testability problems in a class diagram. If there were no client in the software there would be no defined set of executions and thus nothing to test. Thus, after unit testing, failures should only occur because of a misuse due to wrong interactions between objects. These interactions are made more complex if client-supplier dependencies traverse inheritance trees.

The cost and difficulties of testing object oriented systems can be reduced by the basic factors of design principles and planning. Presence of such factors that may generally lead to better testability includes:

- Representation and Implementation characteristics;
- Built-in capability;
- Test case design and information associated with it;
- The environment to support the test; and
- Software testing process.

## 2.2 TESTABILITY ARTIFACTS

Software testing is one of the highly expensive and time-consuming activities of software development life cycle. It has been observed that about 40 to 50 percent of the overall development budget is spent on software testing. Much more resources are required during the activities of test design, test preparation, test execution, and test result analysis. Certain characteristics of a software system make it easier or harder to test it and to analyze the test results. Testability comprises these characteristics and it is an important factor to assure an efficient and effective test process. Testability is often overlooked as a quality attribute of the architecture because it is not a quality directly visible to users.

Voas et al defines testability of a program P to be a prediction of the probability of software failure occurring if the software were to contain a fault, given that software execution is with respect to a particular input distribution. This definition relates the testability with the ability to make the software fail and to the context within which the software is operated. Other definitions of testability relate testability to have visibility into the software, being able to control the software, and being able to detect failures. Finally, testability is related to the size or complexity of the system as it relates to the effort needed to implement test criteria.

Testability is a requisite quality attribute of software with the intent to maintain or improve the level of aggregate quality. Testability is the degree to which a software artifact facilitates testing in a given test context. It is directly related to test effort reduction. A lack of testability, like other design faults, is expensive to repair when detected late during software development life cycle. Therefore, testability should be addressed already during reviews of early development artifacts.

A systematic approach to high testability of software system is called testability engineering. Testability engineering makes the definition of effective test coverage criteria. Testability engineering also makes the test automation easier, and tackles the oracle problem. The references of testability in the literature relate it to the ability to observe the internal state of the software under test and to control that state in order to begin tests at various points in the software's state machine. Therefore, to increase testability, the observability of the software and/or its controllability must be improved. Some definitions also discuss it as being able to detect the failures.

Testability reviews are one of the techniques of testability engineering that can be applied throughout the software development phases. They allow test engineers to evaluate the testability of software artifacts early and to trigger necessary design changes when it is still relatively inexpensive. Testability can be checked as part of reviews with a broader perspective or as pure testability reviews with testability considerations as their main focus. The rigidity of the review process can range from informal to formal.

No doubt software testability is most obviously relevant during testing phase, but paying attention to testability early in the development life cycle may lead to significant improvement in testing phase and potential improvement in testing effort. This may lead to quality end software, as testability is one of the key factors to software quality.

There is no absolute definition and measure of software quality. Software quality may always define in the context of particular customer needs. Similar can be said for software testability. Whether a component or system is testable or not depends on the context, which includes test constraints, intended use of the component, test criteria, and test tools. These are described briefly in the following section.

### 2.2.1 Test Constraints

Cost and quality are two basic testability constraints in testing. A comprehensive testing process consists of activities that identify all potential resource costs. Test costs depend on the efficiency of the test activities, while the quality achieved depends on their effectiveness. These test constraints have to be taken into account when estimating software testability. This is because of the problems caused by lack of testability, which primarily affects the test effort and the fault detection rate of tests based on particular test criteria. Additional test constraints relevant to testability may be test duration or test automation feasibility.

### 2.2.2 Intended Use of the Component

A component to be tested can be intended for a particular use within a system or for reuse within different systems. If a component to be tested is intended for a particular use within the system, the usage context of the component may be well defined and can be used as a basis for test case definition. On the other hand, for reuse within different systems, there may be a problem that test results may be invalid if the component is executed within a new environment not known beforehand. A test approach differs for different types of intended use and so do the testability requirements.

### 2.2.3 Test Criteria

Different issues of design for testability should be addressed depending on the test criteria applied. Efficiency and effectiveness are less sensible to characteristics of the program code when specification based test criterions are used.

### 2.2.4 Test Tools

With the trend of automating the task, there are many software tools, which may be used to aid testing process. Such tools may help to achieve accurate results, in reduced time and effort. The availability and capability of suitable test tools often leads to enhanced level of testability of particular software.

## 2.3 TESTABILITY FACILITATORS

Testability has been defined by ISO as one of the attributes of software. It focuses on the effort needed to validate the software product. Several characteristics have been identified in literatures that contribute to a software system's testability. Figure 2.3 shows the Binder's fish-bone model. Therefore, there is a range of typical factors that are expected to facilitate testability. Such facilitators have been identified, which helps in assessing the level of testability. These are classified as *Internal*, *External*, and *Environmental Facilitators*, and described briefly as follows:

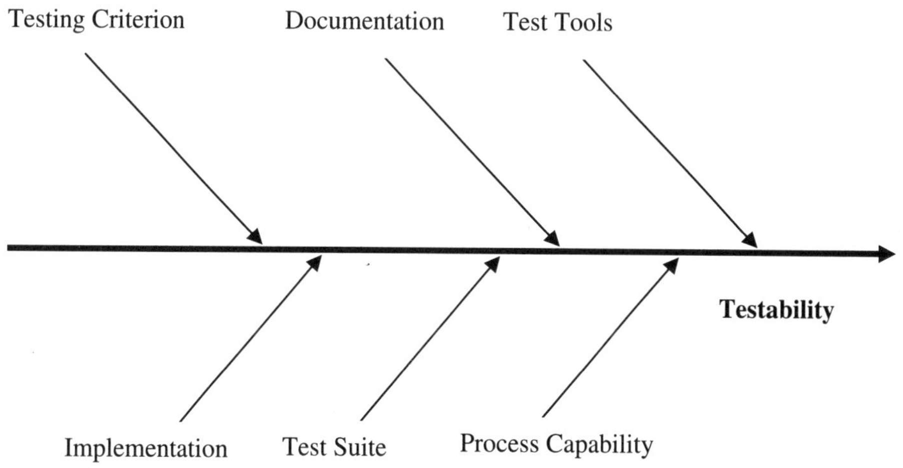


**Figure 2.3: Testability Fish-Bone Model**

### 2.3.1 Internal Facilitators

James Bach defines testability as 'how easily a program can be tested?'. The following internal facilitators to software testability have been identified and depicted in figure 2.3.1.

**Operability:** It refers to the ability of the software to be operable or functional with given setup and environment. This becomes one of the fundamental characteristics of testability, as make feasible the execution of the software and hence test-run.

**Controllability:** It refers to a measure of 'how easily a tester or a user can create difficult scenarios to test the software under extreme circumstances'. For example, behavior of an application cannot be tested very easily when the hard disk is full or table overflow conditions exist.

**Simplicity:** It refers to the general case of construct usages in software. Several types of program simplicity have been identified including *functional simplicity, structural simplicity,* and *code simplicity* described briefly as follows.

- Functional Simplicity: The minimum feature set that is necessary to meet requirements.
- Structural Simplicity: The ease of understanding with module internal interactions coupling.
- Code Simplicity: The ease of facilitative aspects of code for inspection, review, walkthrough purposes.

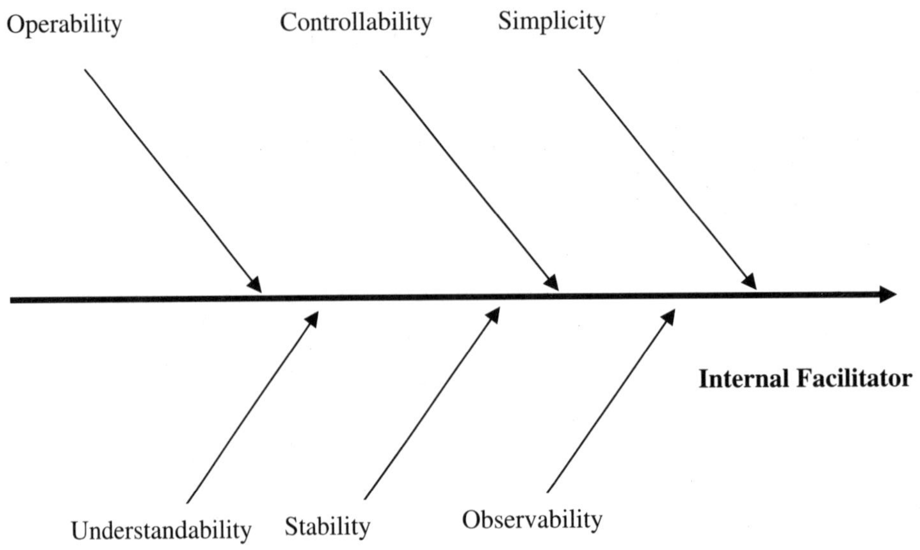


**Figure 2.3.1: Internal Facilitators for Software Testability**

**Stability:** Changes to the software are infrequent, controlled and communicated. If so, such changes to the software should not invalidate automated tests.

**Understandability:** It refers to the ease of understanding and depends on two factors. First is the availability of artifacts for users, such as function specifications, interfaces, programs, testing documents. Second is the availability of artifacts for developers, such as user manual and reference documents.

**Observability:** It refers to the ability of the test tool to capture data and information from the system under test to allow the tester to compare results with actual. This may include hidden information not normally seen by a manual tester.

### 2.3.2 External Facilitators

With the controllable and observable input-output, compliance can be decided. Other facets of testability have been proposed in the literature that influence the ease of testing include the following:

- The amount of difficulty in setting up drivers to execute a component and creating stubs for functionality that does not yet exist,
- The complexity and amount of inherited, parameterized, and polymorphic types in the software, and
- The thoroughness of specification and design information available for test design.

External facilitators to software testability has been shown in figure 2.3.2, and described in the following section.

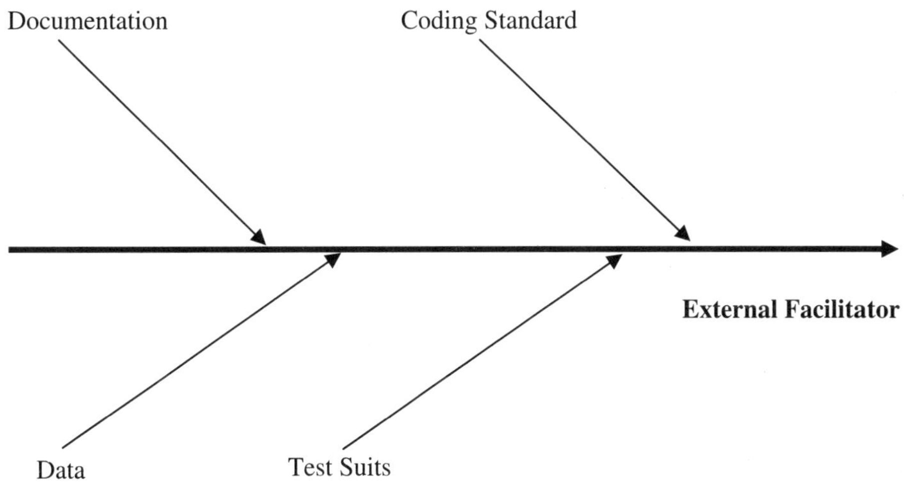


**Figure 2.3.2: External Facilitators for Software Testability**

**Documentation:** Correct and complete requirements specification documentation is of prime importance in software testing phase, because it will assist to identify the required and specified behavior. Documentation process requires two resources, two steps and two different skill sets: the subject matter experts understand the application and the functionality to be verified, and the automation engineers usually have a development background and know how to code in the scripting tool. Availability of good quality and comprehensive documentation enhances the testability, being facilitative to test processes, planning and implementations.

**Coding Standard:** Assertions must be used when available in the development language. Any assumptions made by a developer must be subject to assertions. The assertion output must be available at run time through some interface for the tester or test tool to utilize.

**Data:** Testers should be provided copies of the data files used by developers, at the time of testing. Hence, there is a need to check whether the data is being provided with the software under test. These should be critiqued by the test group to provide feedback to the developers for unit testing. Testers should have ownership of the test database and the test environment. At no time should changes to either be done without test team authorization.

**Test Suite:** Generally, a well-organized collection of test cases is called test suite. A test case is a specific set of steps with expected results and various additional pieces of information. A test suite contains more detailed instructions or goals for each collection of test cases including a section where the tester identifies the system configuration used during testing. A test suite that is ready to be executed is called an executable test suite. There exists a test harness that is integrated with the test suite. This integration is in such a way that the test suite and the test harness can work on sufficiently detailed level to communicate with the system under test. Similar to the system-under-test, test suites need documentation detailing the implemented tests, a test plan, test results of previous test runs and reports.

### 2.3.3 Environmental Facilitators

In spite of internal and external facilitators, environmental facilitators do have their own importance in estimating testability of a software system. Environmental facilitators may ultimately decide the extent to which a software system may testable, and hence considered to be a major contributor for assessing software testability. Following environmental facilitators to software testability have been identified, briefly described and shown in figure 2.3.3.

**Test Tools:** It is a major issue to choose a test tool that should match with the tools, environments, and processes used by the developers. Each test tool has strengths and weaknesses. While it is true that any tools can be made to work in any environment against any system, it is also true that this is not always cost effective. The cost of the extra development it takes to make a tool work against a system will be included in the overall cost of the testing. Therefore, it is required to select a tool which gives the best ratio of testing to cost.

**Testing Criterion:** Software project under test must use testing criteria to specify which part of the software is to be tested. Software testing criteria may establish a lower bound on the validity of software and an upper bound on the number of test cases required. Therefore, a balance is required between the verification and required amount of testing effort. Software testing is one of the important methods of software quality, reliability and safety assurance. Importance of the software testing is increasing as a result of the extension of the computer system use in such safety-critical areas as nuclear power engineering, aviation, cosmonautics, and railway transport. Software testing criteria including test data adequacy and coverage criteria, plays an important role in the entire testing process. It is used to measure the test quality when a degree of adequacy is associated with each test set. It also assists in test data selection and different test sets are considered as equivalent if they are satisfying the same criterion.

**Timing Issues:** Critical timing paths should be clearly specified so that tests can be devised for validating them. The test tool should be able to control virtual time in the application. This will be different for each system. It should be more than simply changing the real-time clock.

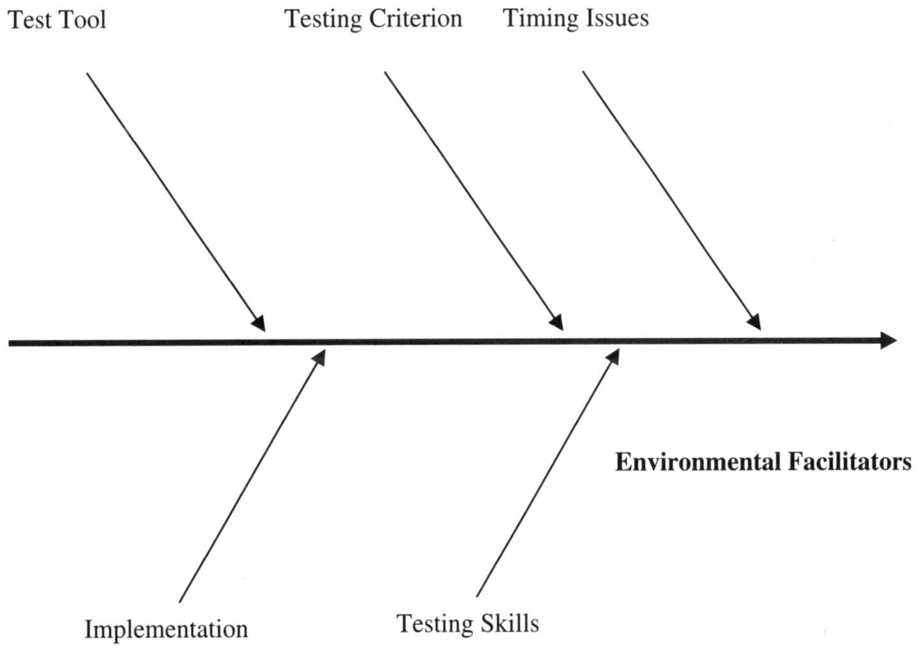


**Figure 2.3.3: Environmental Facilitators for Software Testability**

**Implementation:** Testing requirements vary substantially based on target market, implementation technologies used, development practices and others alike. Test implementation is the testing task of implementing the software testing work products. Its basic objective is to implement and debug the test software. The major part of the implementation of an application consists of source code expressed in one or more programming languages.

**Testing Skills:** The test team must have at least few technically skilled members for understanding technical documentation and interfacing with the development team. The increasing complexity of today's systems requires more technical knowledge of the internal working of systems than in the past. If testers are not adequately skilled in the task, good quality testing cannot be assured.

## 2.4 LEVELS OF TESTABILITY

The one of the major constants of software development is change. The rate of change has accelerated substantially in the last decade; the complexity of applications and environments has increased and schedules have been pinched. Under these conditions, software quality tends to suffer. To cope up with the changes and to improve the software created, there is a need to find ways to optimize Quality Assurance Processes. Improved testability is one of the ways to diminish the widening quality gap.

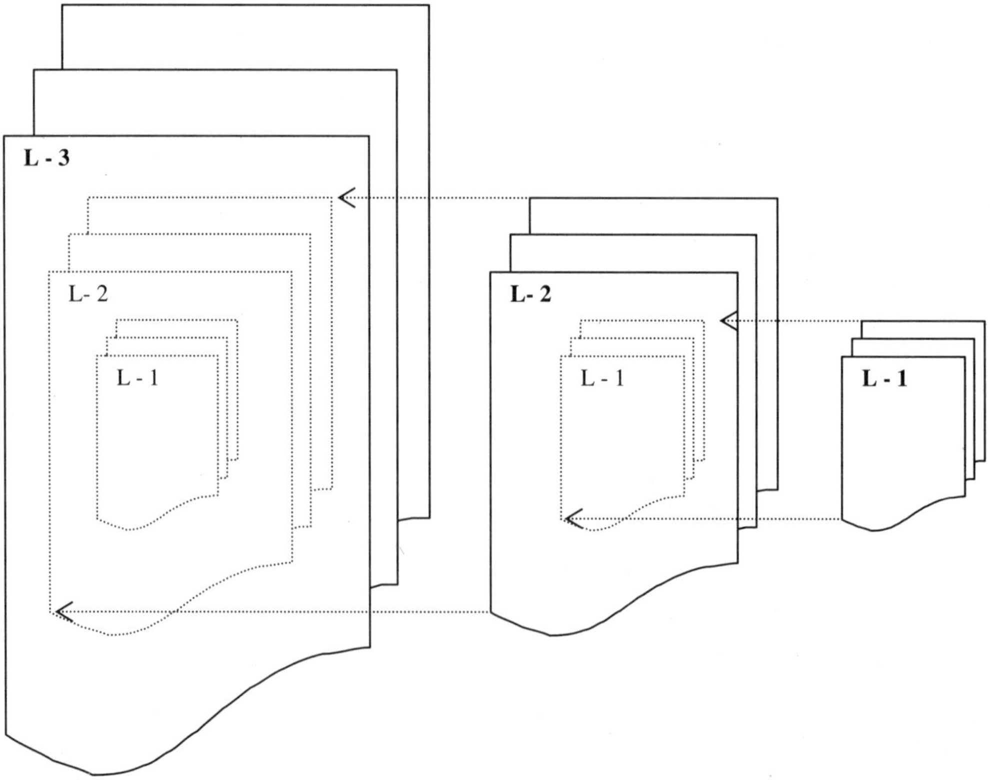


**Figure 2.4 (a): Testability Levels**

In the early days of ad-hoc testing, testing has found a home of its own in many organizations. There was little or no support available to the tester in the way of automated tools. CASE tools proliferated for the development staff, but testers were forced to continue on with manual methods. Whereas testing was primarily performed by programmers as a sort of secondary function in the early 70s, in the late 70s and the mid-80s responsibility for testing resided within a Quality Assurance or Software Quality Assurance group. More recently, the focus of Software Quality Assurance groups has changed into more of a process monitoring function including monitoring of testing, and separate testing groups are proliferating. Testing tools have been growing in both quantity and quality. Planning for testability during the design and coding phases of a software product can provide for significant efficiency in testing the product and has a very small cost and schedule impact on the development processes.

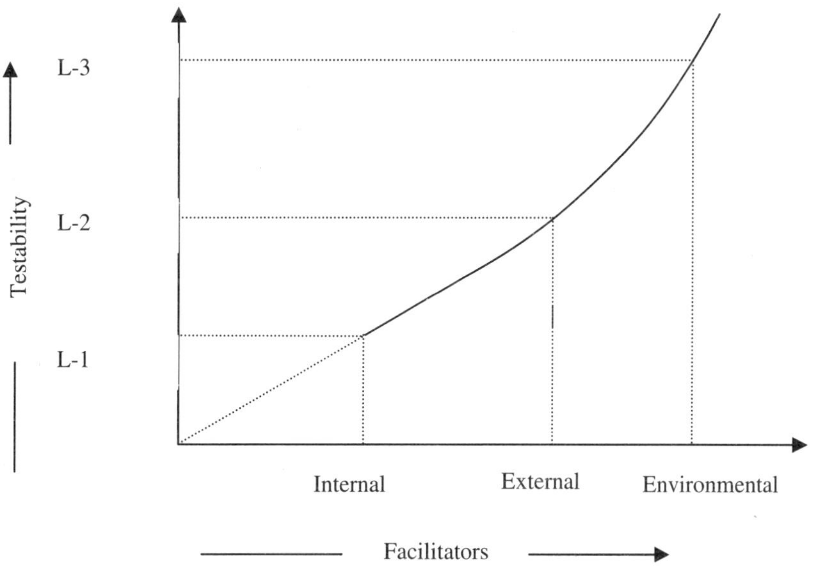


**Figure 2.4(b): Testability Levels and Their Facilitators**

Development tool vendors are always upgrading their tool sets. Developers do not always use best practices when using these development tool sets. Application Programming Interface (API) differences in seemingly similar operating systems may cause unexpected behaviors in scripts. These are some of the factors that can limit the effectiveness of test automation tools. Far too often, a sizable fraction of the cost of automating testing is simply trying to get the tool to coexist with the application. Commonly called 'building workarounds', the act of programming in the test tool language to make up for shortcomings in the tool and eccentricities in the application is an expensive, difficult-to-maintain process, which limits the amount of testing.

Testability leveling can be seen as a holistic approach to these kinds of problems at a high, organizational level. Figure 2.4 (a) & (b) depicts the levels of software testability, and its correlation with its facilitators, while table 2.4 shows the testability levels, testability types and facilitators checklist.

**Table 2.4: Testability Levels, Testability Types and Facilitators Checklist**

| Testability Levels | Testability | Facilitators | | |
|---|---|---|---|---|
| | | Internal | External | Environmental |
| L – 1 | Low | √ | | |
| L – 2 | Medium | √ | √ | |
| L – 3 | High | √ | √ | √ |

**Testability L-1:** For a software system to be L-1 testable, it should qualify at its internal facilitators. If a software system is of testability L-1, it may be regarded as low testable.

**Testability L-2:** For a software system to be L-2 testable, it should qualify at its internal and external facilitators both. If a software system is of testability L-2, it may be regarded as medium testable.

**Testability L-3:** For a software system to be L-3 testable, it should qualify at its internal, external, and environmental facilitators. If a software system is L-3 testable, it may be regarded as highly testable. Moreover, this is the desired level of testability for software systems to be tested adequately, effectively and efficiently.

## 2.5 TESTABILITY ESTIMATION

In order to estimate the testability, quality attributes needs to be determined. The effort needed for testing a particular piece of software is measured through testability of software. It has been proved that software testability is a key factor to software quality, as it determines the effectiveness of test and so the likely correctness of the program. Therefore, measuring testability is a challenging and most important task towards estimating testing effort. One of the most common strategies used for measuring testability is with respect to control flow test strategy because of the following facts:

- Design phase can maximize the testability.
- It can improve test coverage within limited resources.
- Testability measures can also be used for predictions to provide support to management, as testing consumes much time and effort in the development process.

In addition, during testability measurement, it is required to identify software characteristics, which contribute to this quality attribute. Testability can also be defined for the control flow based test strategies as the number of paths required by the strategy. More precisely, testability is defined as the minimum number of test paths to provide total test coverage, assuming that such coverage is possible. Testability with respect to control-flow-based strategies can be determined by measuring the flow graph. This is a well-researched area, and a common framework exists for defining all flow graph metrics. Therefore, flow graph is considered as one of the measuring tools to estimate software testability, in particular context.

Black box testing and glass box testing, are the two well-known strategies for selecting test data. For black box testing, the test data is derived from the specification, analyzing in particular the specified function and its partial functions, the domain and the sub domains, and the function's range. For glass box testing, the test data is derived from the program analyzing the program paths, the data flow and the expressions and data values, that is, the program states. This strategy is defined either by the control flow, the data flow or the states of a program. Although all of these strategies are worthwhile and have individual merits, concentrate on control flow strategies. Followings steps are recommended while testing the software using control flow testing strategy:

- Traverse every node at least once.
- Trace through every edge at least once.
- Trace every path at least once.
- Trace at least once around each loop.

Several concepts and approaches to evaluate different aspects of software testability have been proposed. Some are listed and discussed in the following section.

**Complexity:** It is one of the key aspects to software testability. Several studies have been done to use this aspect to measure the testability of software. A selection from the guidelines concluded from such studies is briefly stated as follows:

- Use complexity of the control flow to estimate the effort necessary to achieve white box test coverage.
- Use data flow analysis to estimate the effort necessary to achieve test coverage related to a set of data flow testing criteria.
- Take care of cohesion, as cohesion is an indicator of class complexity and hence test complexity.

**Coupling:** Coupling is another important aspect to software testability, and has been used in estimating software testability by various researchers and practitioners. Some of the most relevant studies have concluded with valuable contributions, including ones stated below:

- Use testability metric while estimating software testability.
- Use testability metric related to dependencies to other systems, at the time of estimation.

**Object oriented software testability:** It has two key facets- observability and controllability. Several studies have been done on observability and controllability to predict software testability. A selection from the guidelines concluded from these studies are as follows:

- Use domain-to-range-ratio to indicate observability by comparing the size of the output range of a component to the size of its input domain.
- Use mutation analysis based technique to estimate the observability for each statement in a program in terms of the probability that a fault in a statement may remain undetected, given a particular input distribution..
- Use metric related to the observability of test progress.

Estimating program testability remains an open fundamental problem. An experimental technique based on seeding simple faults, affecting individual locations in the source code has been described by Voas, which is untrustworthy for complex software. Voas testability measurement estimates a minimal fault size. Estimating testability requires knowledge about population faults and the effect of input distribution. A program is said to have high testability if it tends to expose faults during random black box testing, producing failures for most of the inputs that execute a fault. A program has low testability if it tends to protect faults from detection during random black box testing, producing correct output for most of the inputs that execute a fault. An input distribution is the distribution of probabilities that elements of the domain are selected. Once inputs are selected, the program is then executed on these inputs and the outputs are compared against the correct outputs.

## 2.6 TESTABILITY ANALYSIS

Software testability analysis is a function of a program, input selection criteria pair, and measures the benefit provided by the software-testing scheme to a particular program. Several ways with different perspective are there to define the benefit of tests and testing scheme. For example, if software testability has been referred to as the ease with which inputs can be selected to satisfy structural testing criteria with a given program, and if it were difficult to find inputs that satisfied structural coverage criteria. Then, with this perspective, that program is said to have low testability with respect to that coverage criteria. Another way of defining program testability is as a prediction of the probability that, for the given input selection criteria, existing faults will be detected during testing. In this case software testability will be regarded as the way of predicting whether a program would reveal existing faults during testing according to the given input selection criteria.

The first reference for testability analysis emphasizes more on exercising software structure during testing. It implicitly assumes that the more software structure exercised during testing, greater will be the likelihood that existing faults will be revealed. Therefore, any straight-line program without any conditional expression or loop would be assigned a higher testability than any program with a more complex flow of control. While on the other hand, whenever, testability is defined to be based on the probability of fault detection, a straight-line program without any conditional expressions or loops could potentially be assigned a lower testability than a program with more complex flow of control.

Different test cases should be generated for different input selection criteria. Random black box method may be used to select the input, and the selection may be dependent upon the structure of the program, or may be based on other data or testers' intuition. Voas's definition indicates that the testability of the software is strictly related to the ability of the software to hide faults during testing when the inputs are selected according to some input selection scheme.

Another reference to testability is semantic based definition. In order to have greater testability, software must return failure if a fault exists. Software failure is nothing but an incorrect output that was caused by a flaw in the program, not an incorrect output caused by a problem with the environment or input on which the program is executing. A fault, by definition, is a structural imperfection in a software system that may lead to the system's eventually failing. This semantic based definition of testability predicts the probability that tests will uncover faults if a fault exists. Software is said to have high testability for a set of tests, if the tests are likely to detect any existing fault. On the other hand, if the set of tests are unlikely to detect any fault that exists, the software is said to be of low testability. To predict such type of probability, a testability analysis technique should be able to quantify whether a fault will be executed/uncovered or not.

### 2.6.1 Designing for Testability

Testability can be viewed as having a negative impact of design quality overall. For example, increasing visibility of attributes decreases the information hiding quality of the design and may threaten the modularity of the design. Reporting 100 percent visibility for the attributes in a piece of software under test would seem to completely defeat information hiding. However, if that level of testability is gained by adding a separate test interface that gives explicit access to attributes, the impact on information hiding is controlled and focused in a single interface.

Tool support can be provided to ensure that inappropriate – any production code using the test interface – uses of the test interface are identified and removed. However, the test interface, and its implementation, increases the size of the executable. This has a negative impact on resource-constrained devices such as embedded systems or wireless sensor networks. There may be process-oriented solutions to this problem such as having the test interface present in the code during testing but removed during product building. There are obvious problems with this proposal as well.

### 2.6.2 Enhancing Testability

Testing is a major cost driver during development and maintenance of software. The overall effort spent on testing not only depends on human factors, process issues, test techniques, and test tools, but also on characteristics of the software development artifacts. The degree to which a software artifact facilitates test tasks in a given test context is generally referred to as testability. Testability is an important quality characteristic of software. A lack of testability contributes to a higher test and maintenance effort. Metrics can be used to locate parts of a program, which contribute to a lack of testability.

improve testability it is required to identify those phases/parts of software that lack testability. This can be done by reviewing software artifacts or with the help of metrics based on design documents and source code. Several approaches and guidelines have been proposed to help to develop testable software from the beginning or to improve testability. Beizer proposed tips on 'how to develop testable code with respect to test techniques such as path testing'.

Similarly Jungmayr et al. presented a guideline on how to improve testability by reducing coupling. Further studies show that Test-First-Design reduces coupling by developing test code before implementing functionality. Vanderwall et al. provided a guideline on how to avoid test automation problems, while Gao proposed guidelines on how to design component-based software for testability.

## 2.7 INCORPORATING TESTABILITY

Testability is one of the major factors determining the time and effort needed to test a software system. It is costly to redesign a system during implementation or maintenance in order to overcome the lack of testability. Testability is not a technological innovation. It is a mindset that creates a constant awareness of the importance of ease-of-testing during different phases, for example, analysis, design, coding and testing. It is no doubt critical to the development process.

A product that cannot be readily tested is not really manufacturable. Therefore, it is important to deploy testability into the system continuously right from the beginning. Just like testing should occur from the very beginning of a project, project artifacts should be reviewed for testability from the beginning as well. Testability is an important characteristic of a software system, which has to be considered during all phases and activities of software development life cycle. Testability life cycle, shown in figure 2.7, and explained briefly as follows, incorporates testability estimation and assurance at various stages of the software development life cycle.

### 2.7.1 Specification Phase

Testing team will understand the requirement, and its several functionalities. The behavior under abnormal scenario may also be documented along with the requirements of the end users specifying expected outcome. Guidelines may be prepared to assist in this phase including the following:

- Clearly define testable requirements.
- Clearly define measurable requirements.
- Provide well-defined facilitating requirements for software testing.
- Review and evaluate system/component requirements to make sure they are testable and measurable.

### 2.7.2 Design Phase

Checklist prepared in the specification phase should be used to elucidate the system path clearly. This may help testers to decide programs under test with their specific scenario. There are large numbers of software-related factors that effect testability during design. Properly managed, testability may heighten the assurance of product quality and smoothes production scheduling. Guidelines may be prepared to assist in this phase including the following:

- Conduct design for testability of software and components, that is, come out architecture model and interfaces increasing testability.
- Define design patterns, standards, and framework for testable components.
- Review and evaluate software design concerning software testability.

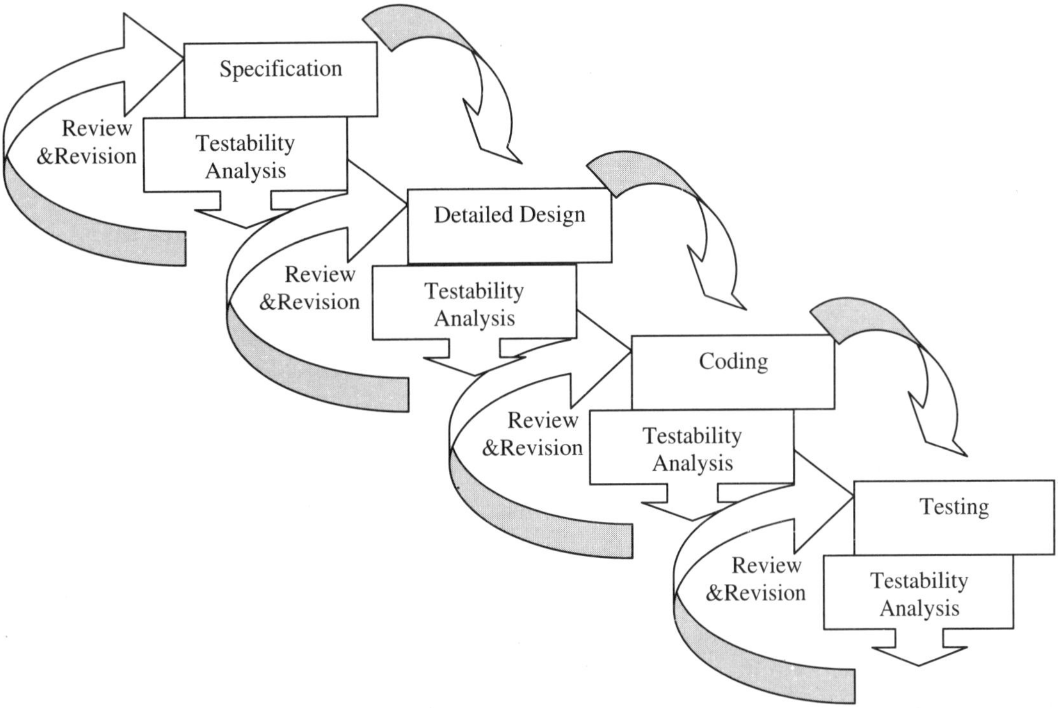


**Figure 2.7: Software Testability Life Cycle**

### 2.7.3 Coding Phase

This is the most critical phase of the development life cycle, where the testability approaches that have been taken in the previous two phases can be incorporated. The documented abnormal scenario in specification phase covering what the application should and should not do may be incorporated and unit tested in this phase. Guidelines may be prepared to assist in this phase including the following:

- Implement testable components and built-in tests.
- Generate software testing facilities and reusable framework.
- Review program/component code based on well-defined testability standards.

### 2.7.4 Testing Phase

All the testability measures taken in the previous phases may be covered by the test plan and test scripts designed in this phase. All the measures should be tested thoroughly along with other functional and non-functional requirements. Guidelines may be prepared to assist in this phase including the following:

- Define achievable component test criteria and high quality tests/scripts.
- Develop, set-up and use test beds and facilities for components.
- Perform component tests and monitor coverage based on defined criteria.
- Verify, evaluate, and measure component testability.

## 2.8 TESTABILITY AND OBJECT- ORIENTED SOFTWARE QUALITY

Quality assurance can be seen from both constructive and analytic aspects. The constructive aspect considers the question of how to build a program of a certain quality, whereas the analytic deals with the question of how to check whether the required quality level has been achieved. The most used technique for analytical quality assurance is program testing, and much time and effort in the software development process is consumed by this task. Thus, it would be useful to have a means for correlating testability with quality factors, of object-oriented software.

### 2.8.1 Quality Factors

A Quality Model for Object-Oriented Design QMOOD has specified 'functionality, reliability, efficiency, usability, maintainability and portability' as the initial set of quality attributes. After reviewing their contributions towards defining design quality, reliability and usability were excluded from the set due to their obvious inclination toward implementation rather than design. Portability is replaced by extendibility. Efficiency was replaced by effectiveness. Maintainability is replaced by understandability.

The goal in adapting an object-oriented approach for design and implementation is to develop reliable, adaptable and flexible software system. This justified the inclusion of reliability as an important attribute of object-oriented design quality assessment. Flexibility is also included as an important quality attribute. Hence the set reduces to Functionality, Effectiveness, Understandability, Extendibility, Reusability, and Flexibility. Software

Assurance Technology Center, SATC, has proposed five of quality differentiators/attributes for the coding and design phase. These are Efficiency, Complexity, Understandability, Reusability, and Testability / Maintainability. It is evident from the discussion that no universally agreed-upon definition for each of high-level quality attributes exists.

### 2.8.2 Design Characteristics

Object-oriented software differs from structured software in terms of its abstraction and real world modeling concepts that take the form of object-oriented design constructs. A fundamental constraint of object- oriented modeling and design is the object, which combines both data structure and behavior in a single entity. Jagdish Bansiya, in his QMOOD model, has discussed the design properties of abstraction, encapsulation, coupling, cohesion, complexity, and design size.

These design properties have been considered as the representative of design quality characteristics in both the structural as well as object-oriented development. Dr. Linda Rosenberg, at SATC (NASA), has described three aspects of object oriented paradigm: Encapsulation, Polymorphism and Inheritance. Polymorphism and Inheritance are two aspects unique to the object oriented approach, while encapsulation in not. Large programs that use information hiding have been found to be easier to modify by a factor of 4 than programs that don't adhere to this technique. Inheritance decreases complexity by reducing the number of operations and operators, but this abstraction of objects can make maintenance and design difficult.

There is no doubt that polymorphism assists the programmer to reduce complexity and improve on many desirable attributes including reusability. Therefore, the three fundamental properties required for an object oriented approach has been considered generally and in the process of development of metrics.

### 2.8.3 Relationship between Software Quality Factors and Characteristics

In order to establish a contextual relationship between design constructs and attributes, the influence of design constructs and quality attributes are being examined (with respect to SATC's attributes). It was observed that each of the design constructs affects certain quality attributes. This is depicted in Figure 2.8.3(a).

Software metrics have been quite an intensive research area for more than two decades, yet its practical applications have been rather limited. Numerous software metrics related to software quality assurance have been proposed in the past and are still being proposed. The current practices of software measurement may be divided into many categories including process metrics, public metrics, fault and failure metrics and product metrics. Process metrics are measures of the software development process, such as overall development time, type of methodology used, or the average level of experience of the programming staff. Fault and failure metrics are the measures of the defect or blemish in object or structure and an unexpected or unwanted event. Product metrics are measures of the software product at any stage of its development, from requirements to installed system.

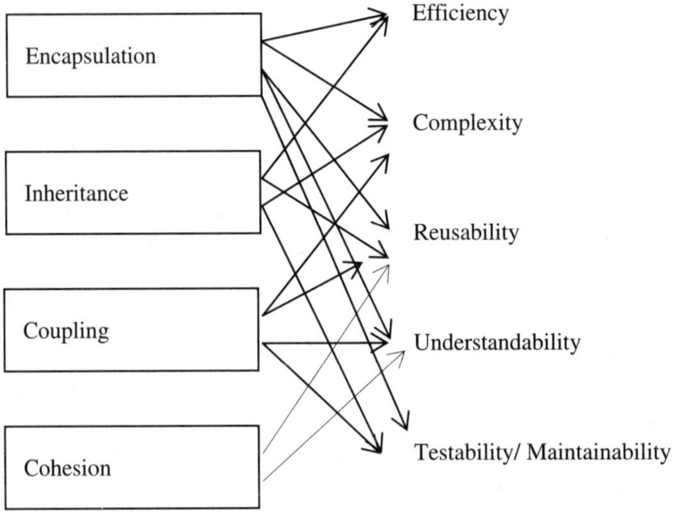


**Figure 2.8.3 (a): Design Constructs and Quality Attributes Relationship**

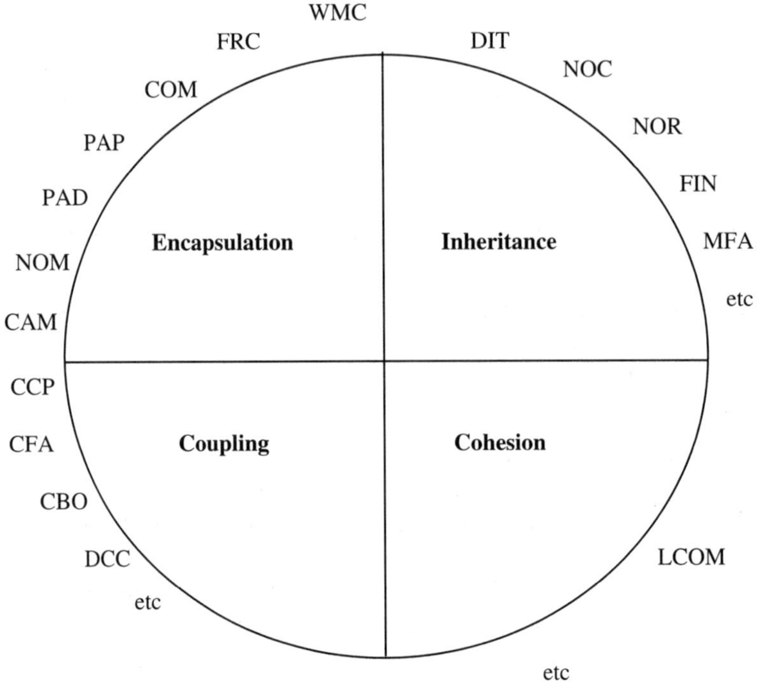


**Figure 2.8.3(b): Metrics and Corresponding Object-Oriented Concepts**

Figure 2.8.3(b) describes a general review of metrics suggested by various researchers/practitioners (MOOD / MOOSE / QMOOD / EMOOSE etc.) and evaluation of object-oriented concepts by these suggested metrics. A critical examination of the existing design metrics revealed that all metrics have relevance with respect to a class, that is, all metrics eventually conduct measures taking class as a basis. This is hardly surprising as 'class' is the fundamental concept of object-oriented software.

## 2.9 CONCLUSION

Software testing is one of the challenging tasks. It is often a very costly development activity of the software development life cycle, which determines whether a program has error or not. Testing is used to assess the compliance of a program to its intended specifications and to assess the reliability of the program to inputs, which were not intended to be part of the specifications. Testability can be defined as the property that measures the ease of testing a piece of code or functionality, or a provision added in software so that test plans and scripts can be executed systematically. Software testability is one of important concepts in design and testing of software program and components. Building programs and components with higher testability always simplifies test operations, reduces test cost and efforts and increases software quality. One way to improve the maintainability of a software system is design for testability, which can address different aspects of software including size, complexity, system structure, built-in-test facilities, distribution, and non-determinism.

Internal state of the software should be controlled to completely test a software component and to assure what has caused a given input. Similarly the output under test should also be observed to assure for the correct processing of given input. Therefore, two key facets have been identified to software testability namely observability and controllability. Several factors have been identified in literature that contributes to a software system's testability. Figure 2.9 depicts software testability integration and related artifacts.

Measuring testability is a challenging and most important task towards estimating testing effort. Testability can also be defined for the control flow-based test strategies as the number of paths required by the strategy. More precisely, testability is defined as the minimum number of test paths to provide total test coverage, assuming that such coverage is possible. With respect to control-flow-based strategies testability can be determined by measuring the flow graph. Estimating testability requires knowledge about population faults and the effects of input distribution.

A program is said to have high testability if it tends to expose faults during random black box testing, producing failures for most of the inputs that execute a fault. Software testability analysis is a function of a program and input selection criteria pair. It measures the benefit provided by the software-testing scheme to a particular program. Several ways with different perspective have been identified to define the benefit of tests and testing scheme. Parts of the program that lacks testability needs to be identified to improve it, which can be done by reviewing software artifacts or with help of metrics based on design documents and source code.

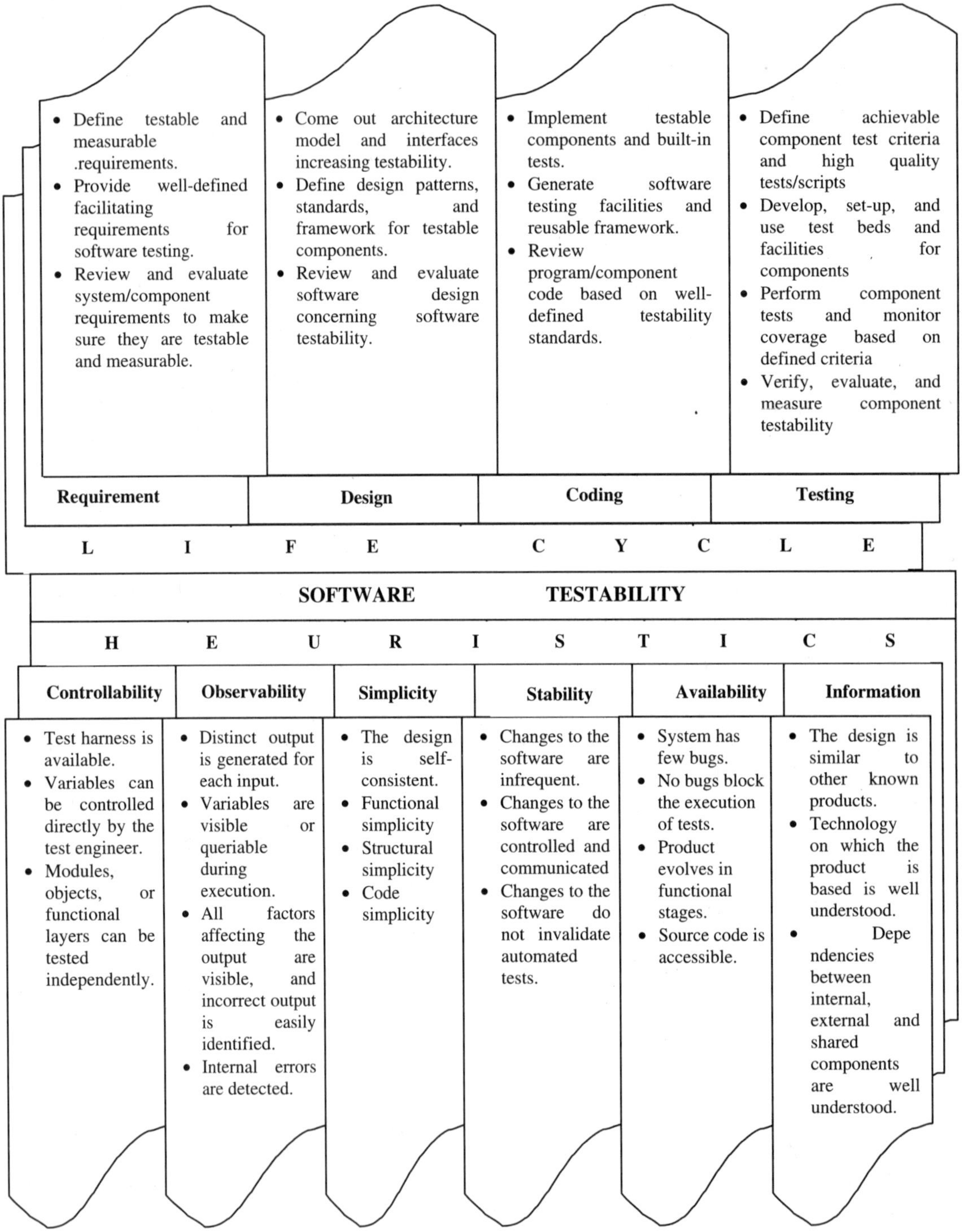


**Figure 2.9: Software Testability Integration and Related Artifacts**

Several approaches and guidelines have been proposed to help to develop testable software from the beginning or to improve testability. A product that cannot be readily tested is not really manufacturable. Therefore, testability is critical to the manufacturing process. There is a common agreement between researchers and practitioners that to deploy testability into the system continually right from the beginning. It is well justified just like testing should occur from the very beginning of a project, project artifacts should be reviewed for testability from the beginning as well.

## KEY TERMS

Testability, Quality factors, Testability factors, Software characteristics, Testability metrics, Observability, Controllability, Testing criteria, Test suite, Test tools, Process capability, Measuring testability, Flow graph, Data flow, Control flow, Error seeding, Complexity, Coupling, Testability analysis, Testability improvement

## REVIEW QUESTIONS

## OBJECTIVE TYPE QUESTIONS

1. _________ enables testers or user to see the external and internal structure of the software.
   (a) Observability
   (b) Controllability
   (c) Both observability and Controllability
   (d) All of the above
2. ________ is the measure of how easily a tester or user can create different scenario to test the software under extreme circumstances.
   (a) Observability
   (b) Controllability
   (c) Both observability and Controllability
   (d) None of the above
3. ___________ phase can maximize testability
   (a) Design phase
   (b) Testing phase
   (c) Coding phase
   (d) None of the above
4. The effort needed to test a particular piece of software can be predicted through_______
   (a) Reliability

(b) Testability
(c) Maintainability
(d) None of the above

5. One of the most common strategies used for measuring testability is with respect to__
(a) Data flow test strategy
(b) Control flow test strategy
(c) Statement coverage
(d) Code coverage

6. For ___________, test data is derived from the specification.
(a) Black box testing
(b) White box testing
(c) Both for black box and white box testing
(d) None of the above

## SHORT ANSWER TYPE QUESTIONS

1. How is software testability different from maintainability? Explain.
2. Find out the major factors that include software testability.
3. What do you mean by level of maintainability?
4. How is software testability related with software maintainability?
5. Summarize various references to testability and identify its facilitators.
6. Distinguish between observability and controllability with an appropriate example.
7. How is testability related with software quality? Explain reasonability.
8. Write down the importance of test suit in establishing test criteria.
9. Elaborate needs and significance of testability metrics in estimating testability of object oriented software.
10. How testability is defined for control flow based strategy?
11. List the approaches to evaluate the major aspects of software testability.
12. What do you mean by seeding simple faults during testability evaluation?
13. What do you mean by testability analysis?
14. Explain the concept of L–1, L–2, and L–3 testability of software. Explore examples for each, and justify their testability.

## DESCRIPTIVE QUESTIONS

1. Review a paper on software testability. Find out the set of factors affecting software testability.
2. How can software testability be estimated in early stage of software development life cycle?
3. Propose a model to estimate the object oriented software testability in design phase.
4. Empirically validate the model proposed in question 3 with pre-try out and try-out.
5. Develop a metric suit used to quantify software testability in early stage of software development life cycle.
6. Validate the proposed metric suit. Implement this suit on any small piece of software and estimate the quantitative value of testability.
7. Find out a set of testability factors for object oriented software and another set of object oriented software characteristics. Establish a relationship between these two sets.
8. Develop a framework to predict appropriate software testability.
9. 'Software testability is key factor to software quality.' Justify.
10. Suggest few guidelines for improving software testability.

## REFERENCES


1. Adams, Ed (2001). "Achieving Quality by Design, Part II: Using UML." The Rational Edge, September.
2. Bach, James (1999). "Heuristics of Software Testability."
3. Benner, Joe (1999). "Understanding Microsoft Active Accessibility", in Visual Test 6 Bible by Thomas R. Arnold II, IDG Books.
4. Bray, Brandon (2002). "Compiler Security Checks In Depth," MSDN, February,
5. Cohen, Noam (2000). "Building a Testpoint Framework," Dr. Dobb's Journal, March.
6. DiMaggio, Len (2000). "Looking Under the Hood," STQE Magazine, January.
7. Binder, R.V. Testing Object-Oriented Programs: A Survey. RBSC-94-002. Robert Binder Systems Consulting, Inc., Chicago, 1994.
8. Binder, R.V. The FREE A#roach to Testing Object-Oriented Systems. RBSC- 94-003, Robert
9. Ould, M.A., and Unwin, C. Testing in Software Development. Cambridge University Press, Cambridge, UK, 1986.

10. Foote, Brian and Joseph Yoder (2000). "Big Ball of Mud" in Pattern Languages of Program Design 4, Harrison et al, eds. Addison-Wesley.
11. Fowler, Martin (2000). Refactoring, Addison-Wesley.
12. Fowler, Martin (2000a). UML Distilled, Addision-Wesley.
13. Fowler, Martin (2001). "Separating User Interface Code," IEEE Software, March.
14. Friedman, Michael & Jeffrey Voas (1995). Software Assessment: Reliability, Safety, Testability, Wiley.
15. Gelperin, David (1999). "Assessing Support for Test," Software Quality Engineering Report, January.
16. Guckenheimer, Sam (2000). "What Every Tester Should Know About UML," STAR East, May.
17. R. V. Binder, "Design for Testability in Object-Oriented Systems," *Communications of the ACM*, vol. 37, pp. 87-101, Sept. 1994.
18. L. Briand, J. Daly and J. Wuest, "*A unified framework for coupling measurement in object-oriented systems*," Fraunhofer Institute for Experimental Software Engineering, Germany, Tech. Rep. ISERN-96-14, 1996.
19. L. C. Briand, Y. Labiche, and Y. Wang, "*Revisiting strategies for ordering class integration testing in the presence of dependency cycles*," Carleton University, Ottawa, Canada, Tech. Rep. TR SCE-01-02, 2001.
20. P. Eades, X. Lin, and W. F. Smyth, "A fast & effective heuristic for the feedback arc set problem," *Information Processing Letter*, no. 47, pp. 319-323, 1993.
21. M. Fowler. *Refactoring - Improving the design of existing dode*. Addison-Wesley, Object Technology Series, 1999.
22. Binder Systems Consulting, Inc., Chicago, 1994. Cox, B.J. Planning the software industrial revolution. IEEE Softw. Nov. 1990, 25-33.
23. Hoffman, D. Hardware testing and software 1Cs. Proceedings, Northwest Software QuaSty Conference (Portland Ore., Sept. 1989), pp 234-22.
24. Hoffman, D. and Strooper, P. A case study in class testing. In Proceedings of CASCON 93. IBM Toronto Laboratory, Oct. 1993, pp 472-482.
25. Hayes, Linda (2000). "I Want My Test API," Datamation, December.
26. Houlihan, Paul (2002). "Targeted Fault Insertion," STQE Magazine, May.
27. IEEE (1990). IEEE Standard Glossary of Software Engineering Terminology, ANSI/IEEE Std 610.12-1990.
28. Information Systems Examination Board (ISEB) (1999). Software Testing Foundation Syllabus, version 2.0, February.
29. InstallShield (1998). "Creating a Silent Installation."
30. Johnson, Karen (2001). "Mining Gold from Server Logs," STQE Magazine, January.

31. Kaner, Cem (1998). "Avoiding Shelfware: A Manager's View of Automated GUI Testing."
32. Kaner, Cem (2000). "Architectures of Test Automation."
33. Lemos, Robert (2001). "Passwords don't protect Palm data, security firm warns," CNET News.com, March 2.
34. Lesuer, Brian (2001). "Supporting Custom Controls with SilkTest," Test Automation Conference, August.
35. ANSI/IEEE Standard 830-1 984: Standard for Software Requirements Specifications. The Institute of Electrical and Electronic Engineers, New York, 1984.
36. ANSI/IEEE Standard 829-1983: IEEE Standard for Software Test Documentation. The Institute of Electrical and Electronic Engineers, New York, 1987.
37. ANSI/IEEE Standard 1016-1 987: IEEE Recommended Practice for Software Design Descriptiolw. The Institute of Electrical and Electronic Engineers, New York, 1987.
38. ANSI/IEEE Standard 1149.1-1 990: IEEE Standard Test Access Port and Boundary-Scan Architecture. The Institute of Electrical and Electronic Engineers, New York, 1990.
39. R. S. Freedman, "Testability of software components," *IEEE Transactions on Software Engineering*, vol. 17, pp. 533-564, June 1991.
40. E. Gamma, R. Helm, R. Johnson, and J. Vlissides. *Design patterns: Elements of reusable object-oriented software*. Addison Wesley. October 1994.
41. ISO/IEC 9126: Software Engineering - Product quality [Jung99] S. Jungmayr, "Reviewing software artifacts for testability," presented at *EuroSTAR '99*, Barcelona, Spain, Nov. 8-12, 1999.
42. K. Karoui, R. Dssouli, "Testability analysis of the communication protocols modeled by relations," Tech. Rep. TR 1050, Universite de Montreal, Department d'Informatique et de Recherche Operationelle, Nov. 1996
43. Beizer, B. Software System Testing and Quality Assurance. Van Nostrand Reinhold, New York, 1984.
44. Jacobson, 1, Christerson, M., Jonsson, I'. a n d Overgaard, G. Oblect-Onentrd .%@a~-r Eqizc~&g. Addison-Wesley, Reading, Mass., 1992.
45. Lieberherr, K.J. and Holland, 1.M. Assuring good s t y l e f o r objectoriented programs. IEEE .So/iw. (Sept. 19X9), pp. 38-49.
46. Meyer, R. Obj&-Or&ted Softural-r Construction. Prentice-Hall, Englewood Cliffs, N.J., 1988.
47. Ogata, K. Modern Control Engineering. Second ed. Prentice-Hall, Englewood Cliffs, N.J., 1990.
48. OO tool aids software testing. The Outlook. Fall 1993, McCabe & Associates, Columbia, Maryland.

49. Marick, Brian (2000). "Using Ring Buffer Logging to Help Find Bugs."
50. Maguire, Steve (1993). Writing Solid Code. Microsoft Press.
51. Mercury Software (2000). WinRunner Advanced Training 6.0: Student Workbook.
52. Meyer, Bertrand (1997), Object Oriented Software Construction.
53. Mitchell, Jamie L (2001). "Testability Engineering," QAI.
54. Mullins, Bill (2000). "When Applications Collide," STQE Magazine, September.
55. Nyman, Noel (2001). Personal communication.
56. Nyman, Noel (2002). Personal communication.
57. Payne, Jeffery, Michael Schatz, and Matthew Schmid (1998). "Implementing Assertions for Java," Dr Dobbs, January.
58. Philips, Todd (2002). "Testing Applications with AppVerifier," MSDN, June.
59. Pettichord, Bret (1999). "Seven Steps to Test Automation Success," Star West, November 1999.
60. Deming, W.E. Out of the Crisis. MIT Center for Advanced Engineering Study, Cambridge, Mass., 1986.
61. Doong, R.K. and Frankl, P. The ASTOOT approach to testing object-oriented programs. Tech. Rep. PUCS-104-91. Polytechnic University, Brooklyn, N.Y, 1991.
62. Freedman, R.S. Testability of software components. IEEE Trans. Softw. Eng. 17, 6, tJune 1991), pp 553-64.
63. J. Lakos. *Large-scale C++ software design*. Addison- Wesley, 1996.
64. J.-C. Lin, S.-W. Lin, and I.-Ho, "An estimated method for software testability measurement," in *Proceedings of the 8th International Workshop on Software Technology and Engineering Practice (STEP '97)*, London, UK, July 14-18, 1997.
65. B. W. N Lo and H. Shi, "A preliminary testability model for object-oriented software," in *Proceedings of the 1998 International Conference Software Engineering: Education and Practice*, Dunedin, New Zealand (M. Purvis, ed.), pp. 330-337, 1998.
66. J. Mason and E. S. Ochotta, "The application of object-oriented design techniques to the evolution of the architecture of a large legacy software system," in *Proceedings of 5th USENIX Conference on Object- Oriented Technologies and Systems* (COOTS '99), San Diego, California, USA, May 3-7, 1999.
67. T. McCabe, "A software complexity measure", *IEEE Transactions of Software Engineering*, vol. 2, no. 4, pp. 308-320, 1976.
68. J. D. McGregor, "Instrumenting for class testing," *Journal of object-oriented programming*, February, 1999.
69. J. Overbeck, "Integration testing for object-oriented software," Ph.D. dissertation, Technical University of Vienna, Austria, 1994.

70. Pettichord, Bret (2000). "Three Keys to Test Automation," Stickyminds.com, December
71. Rubenstein, David (2002). "Who Knows What Evil Lurks in the Hearts of Source Code?," SD
72. Structu Rise (undated). Textract, a screen text capture OCR library.
73. Valaer, Laura and Robert Babb (1997). "Choosing a User Interface Development Tool," IEEE Software, July.
74. Segue Software (1998). Testing Custom Objects: Student Guide.
75. Whittaker, James A (2003). How to Break Software, Addison Wesley.
76. Zambelich, Keith (2001). "Designing for Testability: An Ounce of Prevention," unpublished.
77. Beizer, B. Software Testing Techniques, Second ed. Van Nostrand Reinhold, New York, 1990.
78. Berard, E. Essays on Object-Oriented Software Engineering. Prentice-Hall, Englewood Cliffs, N.J., 1993.
79. Paulk, M.C., Curtis, B., Chrissis, M.B., and Weber, C.V. Capabilio Maturity Model for Software, Version 1.1. CMUJ SEI-93-TR-24. Software Engineering Institute, Pittsburgh, Pa., 1993.
80. Voas, J., Morell, L., and Miller, K. Predicting where faults can hide from testing. IEEE Softw. (March 1991), pp 41-8.
81. Weyuker, E.J. On testing non-testable programs. Comput. J. 1982 25, 4, pp 465-70.

## USEFUL LINKS

http://www.therationaledge.com/content/sep_01/m_qualityByDesign_ea.html

http://msdn.microsoft.com/library/default.asp?url=/library/enus/

dv_vstechart/html/vctchCompilerSecurityChecksInDepth.asp

http://dev.scriptics.com/community/features/Xconq.html

http://www.structurise.com/textract/index.htm

http://msdn.microsoft.com/library/default.asp?url=/library/enus/

dnappcom/html/AppVerifier.asp?frame=true

Times, June 1. http://www.sdtimes.com/news/055/story4.htm

http://www.io.com/~wazmo/papers/seven_steps.html

http://news.com.com/2009-1040-253481.html

http://www1.bcs.org.uk/DocsRepository/00900/913/docs/syllabus.pdf

http://www.laputan.org/mud/mud.html

http://datamation.earthweb.com/dlink.resourcejhtml.

http://support.installshield.com/kb/display.asp?documents_id=101901

http://visibleworkings.com/trace/Documentation/ring-buffer.pdf

http://kaner.com/testarch.html

http://www.stickyminds.com/sitewide.asp?ObjectId=2084&ObjectType=COL&Function=edetail

# Chapter 3

# Static Testing

*Take time to deliberaté, but when the time for action has arrived, stop thinking and go in.*
**Napoleon Bonaparte**

## OBJECTIVES

A successful completion of this chapter will enable the reader for the following:

- To understand and describe static testing strategy in general.
- To appreciate the need, importance and significance of static testing.
- To enumerate and realize numerous static testing principles.
- To list purpose, goals and importance of static testing.
- To present advantages and disadvantages of static testing techniques.
- To understand various manual testing techniques.
- To realize the importance and significance of using manual testing.
- To describe different types of automated testing techniques.
- To visualize the benefits of automated testing.
- To define and understand walkthrough and inspection.
- To realize the facts related to syntax parser and static verification.
- To enumerate and realize generic view of software testing strategy.
- To differentiate static testing from dynamic testing techniques.
- To understand the prescriptive walkthrough process.
- To define the inspection process.
- To realize the importance and significance of technical review in software testing.
- To list different kinds of errors produced as a result of syntax parsing.
- To prepare a technical review checklist.
- To differentiate between managerial review and formal technical review.
- To successfully apply different static testing techniques.

## 3.1 INTRODUCTION

Static testing is performed without program execution, and in an over-broad term. It includes almost everything except conventional program testing by execution. Since dynamic testing requires executable code, static testing can be applied earlier during development. Static testing may be performed manually or even by using special CASE tools. Some kind of the errors, like timing-dependent errors, is hard to find by static testing. Static analyzers are software tools for source text processing. They parse the program text and try to discover potentially erroneous conditions and bring these to the attention of the Verification and Validation team. Static analyzers are very effective as an aid to software inspection and are supplement to, but not a replacement, for inspection.

Static testing is a major process to facilitate the following phase called static analysis. However, these two terms are interchangeably used in literature. Analysis always needs to be performed to reach the conclusion on faults but only after testing has revealed the initial information on unexpected behavior. Static testing may be referred to as the test activities associated with analyzing the products of software development without actually executing the code. It is normally conducted for the purpose of *validating* that a work product such as a design specification properly implements all the system requirements, and *verifying* the quality of the design. It is regarded as one of the most effective means of catching bugs at an early stage of development, thereby saving substantial time and cost of the development. It involves inspections, walkthroughs, and peer reviews of designs, code, and other work products, followed by static analysis to uncover defects in syntax, data structure, and other code components. It may broadly cover anything that can be performed to uncover defects without executing the code.

Static testing may reveal the faulty programming constructs at various levels. Some of the examples are given as follows:

- It enables the checking for variables used before initialization, declared but never used, assigned twice but never used between assignments.
- It looks for possible array bound violation and undeclared variables.
- It analyzes the unreachable codes and unconditional branches into loops.
- It checks for interface fault and will assure for parameter type and number mismatch as well as uncalled function and procedures.
- It checks unassigned pointers and pointer aromatic.

Static testing is particularly valuable when a language, which has weak typing such as C, is used. In this type of the languages, many errors are undetected by the compiler. However, static testing is less cost-effective for languages that have strong type checking like Java, and can, therefore, detect many errors during compilation. A static testing of software is a sound, finite, and approximate calculation of the program's execution semantics. It is cconcerned with analysis of the static system representation to discover problems, and may be supplement by tool-based document and code analysis. Static analysis tools can vet software code, either in source or binary form, in an attempt to identify common implementation level bugs such as buffer overflows. Table 3.1 shows the stages, static testing phases and corresponding checks.

**Table 3.1: Static Testing Stages**

| Stage | Phase | Checks |
|---|---|---|
| 1 | Control Flow Analysis | • Loops with multiple exit or entry points<br>• Unreachable code |
| 2 | Data Use Analysis | • Un-initialized variables<br>• Variables written twice without an intervening assignment<br>• Variables which are declared but never used |
| 3 | Interface Analysis | • Consistency of routine and procedure declarations and their use |
| 4 | Information Flow Analysis | • Dependencies of output variables<br>• Information for code inspection or review |
| 5 | Path Analysis | • Paths through the program and sets out the statements executed in that path |

## 3.2 PRINCIPLES OF STATIC TESTING

Static testing is extremely efficient in highlighting semantic problem areas in the source code regarding misuse of types and variables. A large number of different testing metrics are extracted from code without dynamic execution of source code in case of static analysis technique. On the basis of the measurement values obtained, it is possible to evaluate the robustness and complexity of the source code. Structural metrics are calculated to improve the quality of the code structure, and the code structure can be visualized for review or inspection. All sorts of anomalies in the source code are listed during report generation from the static analysis of the code provided. Compiler only checks the syntax, and hence does not care for these anomalies. While static testing will take care of semantic validity of the source code, and as a result report a large amount of anomalies to developers to judge if changes are required.

A complete static testing may be performed for all the modules in the product before unit testing starts, and after each modification to the unit later in development. A selection of principles of static testing are described as follows:

**Review all the Coding Standard Violations:** All violations of coding standards should be removed. Code standards could be either general good practice, or self-imposed company standards. If portability is an issue, it is advisable to choose a static analyser that specializes in this. If not, consider other static analysers instead, because too many anomalies may be reported that are of no interest.

**Review Complexity of Code:** The complexity of the code must be reviewed. The static analysis reports contain many measures of complexity including lines-of-code, McCabe complexity metric, and depth of loop nesting. In many companies limits on these complexity numbers have been imposed because it is believed that units of larger complexity are relatively more error prone.

**Inspect and Remove Unreachable Code, Unreferenced Procedures, and Unused Variables:** Unreachable code, unreferenced procedures, and unused variables are dead items, and potentially dangerous. Therefore, they should be inspected and removed.

**Report all Types of Data Flow Anomalies:** Whenever variables are used in potential dangerous or suspicious situations, data flow anomalies are reported. There are several types of data flow anomalies including the following:

***A variable is used before it is initialised:*** This is most popular and general anomaly. If a variable is used before it is initialised, it becomes dangerous because it is not known what value the variable will have when it is used. This may lead to a disaster. These types of anomalies must either be removed, or explained and documented.

***A variable is assigned a value and then runs out of scope without being used:*** These cases are not normally so well known by developers. In the majority of cases the anomalies reported of this type do not represent real bugs. All these anomalies shall be examined by the developer, and documented in the code.

***A variable is assigned a value and then reassigned without having been used in between:*** In many cases the use is quite normal: a variable is assigned a value before a loop is entered, and in the loop this variable is reassigned. So again the majority of the anomalies reported of this type do not represent real bugs. All these anomalies shall be examined by the developer, and documented if not located close to each other.

***A parameter mismatch:*** This is nothing but the difference in type of the parameter declaration and the actual variable being passed as a parameter to a procedure. Many compilers perform automatic conversions (casting), which could prove dangerous in certain cases. All parameter mismatches shall either be removed or explained and documented.

***Suspicious casts:*** Suspicious casts appear when a variable is assigned a value that differs in type from the variable's type. Casts are also cases where the types are 'alike' but where the assignment may result in a loss of information. Many compilers perform automatic casting, which could prove dangerous in certain cases. If suspicious casts are unavoidable, use explicit casts, and document them.

## 3.3 STATIC TESTING PERSPECTIVE

Usually the test team receives a vague and high-level requirement document in order to design their tests. Several questions test engineers try to ask, but very often they do not know even what to ask. Similarly, when a developer gets the same vague high-level requirement document, they use it to make assumptions about the missing details. The test team remains uninformed of any changes in the requirements in all cases, and hence the test design does not covers everything in code. In order to resolve the problem, there is a need to know how to capture details during requirement review, design review, and code reviews. This is the point where static testing comes into the picture. Static testing is a form of software testing where the software is not actually used. Instead, methods like code review, inspection and walkthrough are used. Static testing is the analysis of source code without execution. Static testing can be done manually, like inspections as well as be automated by using software tools. Because static testing does not require execution of the code, analysis for errors and vulnerabilities can be done throughout the software development process, and analysis can be conducted across all code paths. It is usually conducted by looking for error signatures or patterns that have cause problems in earlier programs. It may also apply to Black Box testing techniques of requirement review.

When programmers and software managers hear the word testing, the usual thought is the process that follows coding. A kind of testing where the software program is actually running and its performance is tested is called dynamic testing. The opposite of dynamic is static, where the testing is done while the program is not running and is called static testing. But so many people are unfamiliar with static testing. It may be because of the different terminology used like 'Code Inspection' or 'Design Inspection'. Another term very often heard is 'Code Review' or 'Design Review'. Static testing should be started at the right time to get the optimal benefit. When programmers have finished coding the entire system, requirements should be reviewed to help testers design test cases.

### 3.3.1 General Methodology

Identification of a problem in the requirements before it turns into a problem in the system will save time and money. This is the major benefit of static testing. M.E. Fagan wrote in *IBM Systems Journal* that 67 percent of total defects are found during the development in Inspection only while 82 percent of all the defects are found during inspection of design and code. Analysis of source code to expose potential defects is a well-developed branch of software engineering. The typical mode of operation is to make the target code available to a code analysis tool. The tool will then look for a class of problems and flag them as potential candidates for investigation and fixes. There are several benefits in source code analysis including the following:

- It can be done before an executable version of the program exists.
- A certain class of faults, for example memory leaks, is more easily exposed by analysis than by testing.
- A fault is more easily localized, since the symptoms tend to be close to the cause.

Literature review on effective verification process reveals that inspection decreases the cost of detecting a fault by 85 percent. The Fagan diagram shown in figure 3.3.1 captures the importance of static testing. Static testing is extremely efficient in highlighting semantic problem areas in the source code regarding misuse of types and variables. Structural metrics are calculated to improve the quality of the code structure, and the code structure can be visualized for review or inspection. Static testing is a technique whereby a large number of different metrics are extracted from the code without dynamic execution. Based on the measurement values it is possible to evaluate the robustness and complexity of the source code.

Reports from static analysis include all sorts of anomalies in the code. Compilers usually regard such anomalies as legal because they only check the syntax, but static testing will question the semantic validity of the code. Static analyzers normally report a large amount of anomalies, which makes it a tedious task of sorting out which of the anomalies represent real problems. The task of static testing in this respect is to point out potential problems as anomalies and then leave it to the developers to judge if changes are needed.

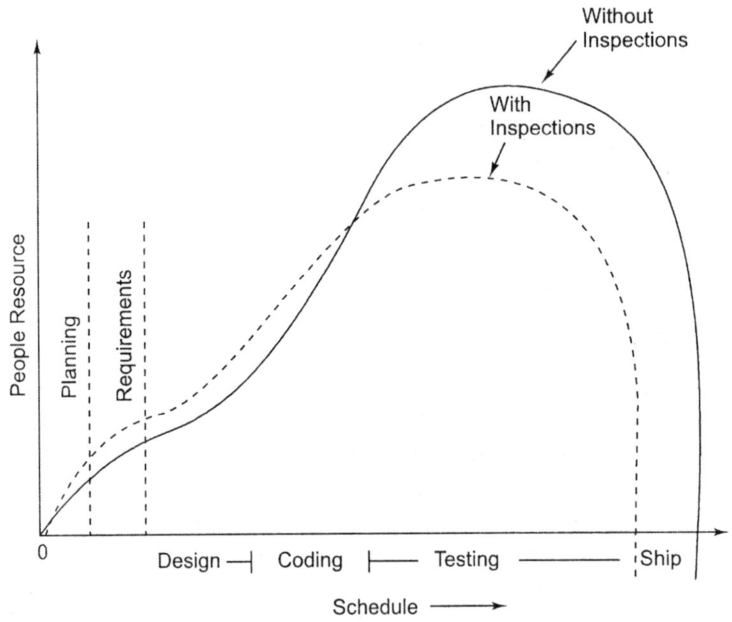


**Figure 3.3.1: Importance of Static Testing**

Static testing is generally not taken as detailed testing, but checks mainly for the sanity of the code/algorithm, and to assess if program is ready for more detailed testing, where methods like code review, inspection and walkthrough are used. There are several advantages of conducting static testing including the following:

- Static testing is a bargain, compared to dynamic testing.
- Static testing is several times more effective.
- Static testing is faster and may achieve 100% coverage.

- Unit cost of detecting bugs by static testing is much lesser than detecting bugs by dynamic testing.
- Static testing is more feasible to perform than dynamic testing.

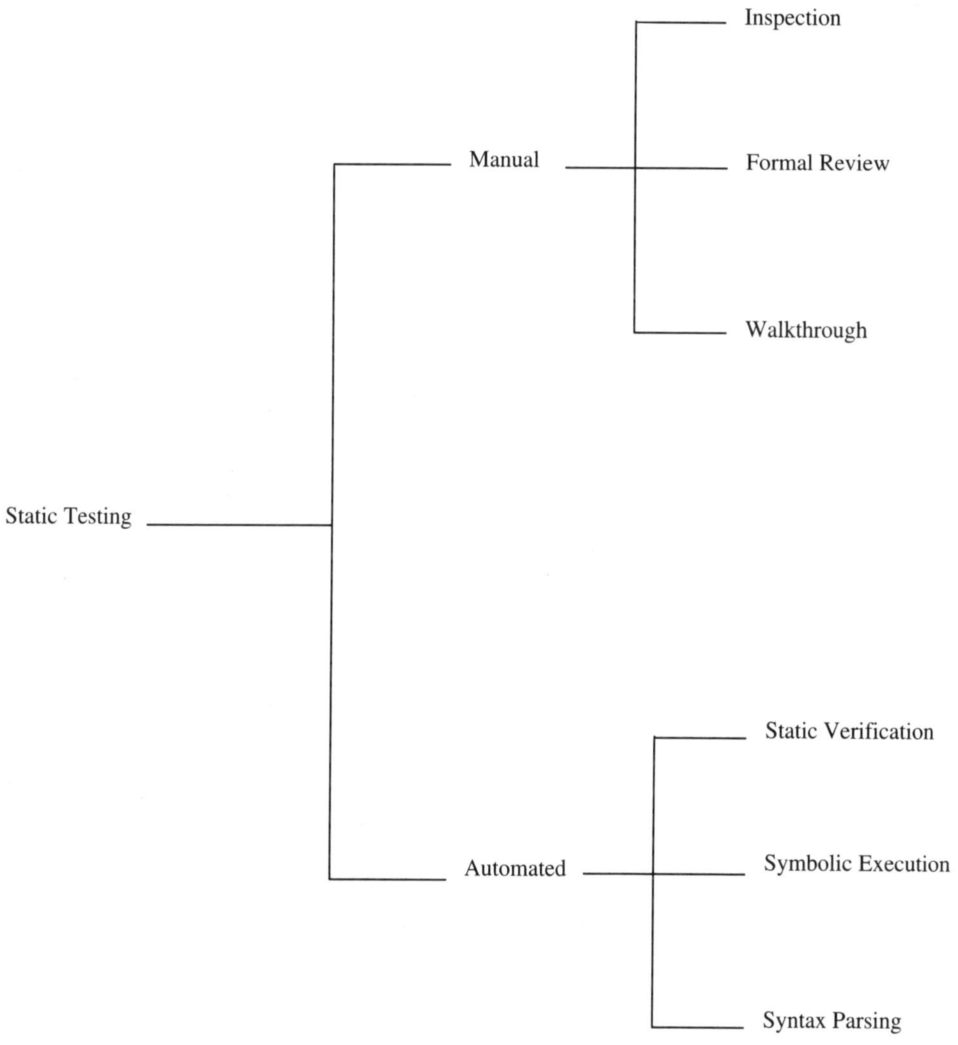


**Figure 3.3.2: Static Techniques**

In spite of several advantages, there are many disadvantages of static testing, reported in the literature. Some of them are listed, particularly in case of using tools, as follows:

- Enormous amount of anomalies are generated by the static analyzer.

- The static analyzers have problems recognizing certain code constructs including function pointers, and identifying variables in condition statements.
- The static analyzers cannot see what goes on outside the unit being analyzed e.g. in the operating system, and libraries.
- Comparison of the old and new report must be performed each time
- It is not feasible to perform static testing after each compilation.
- It is advisable to run static analyzer only before unit testing starts and after modifications to the unit later in the process.

### 3.3.2 A Taxonomy of Static Testing

Static testing does not necessitate the execution of the program. Essential functions of static analysis are checking whether representations and descriptions of software are consistent, noncontradictory or unambiguous. It generally takes care of correct descriptions, specifications and representations of software systems. Therefore, it is a precondition to any further testing exercise. It covers the lexical analysis of the program syntax and investigates and checks the structure and usage of the individual statements. Scope of software testing may be described as:

- Checking the completeness and consistency of program;
- Considering pre-defined rules;
- Comparing the program with its specification or documentation;

Researchers and practitioners are considering that static testing techniques can be performed automatically with the aid of specific tools like parser, data flow analyzers, syntax analyzer, etc. Others also include manual techniques for testing that do not ask for an execution of the program. Figure 3.3.2 depicts the taxonomy of static testing techniques, which are described briefly in following sections:

## 3.4 MANUAL TECHNIQUES

Manual testing is performed by the tester who carries out all the actions on the application manually, step by step. It indicates whether a particular step was accomplished successfully or whether it failed. Manual testing is always a part of any testing effort. It is especially useful in the initial phase of software development, when the software and its user interface are not stable enough, and beginning the automation does not make sense.

Even in the age of short development cycles and automated-test-driven development, manual-testing technique contributes vitally to the software development process. A number of good reasons exist to do manual testing including the following.

- Manual testing allows end users for using prototype software by giving repeatable sets of instructions.
- Manual testing also allows them to be involved early in each development cycle and draws invaluable feedback from them that can prevent surprise application builds that fail to meet real-world usability requirements.
- Manual test scripts give testers something to use while awaiting the construction and debugging of automated scripts.
- Manual test scripts can be used to provide feedback to development teams in the form of a set of repeatable steps that lead to bugs or usability problems.
- Manual test scripts can also form the basis for help or tutorial files for the application under test, if done thoroughly.
- Manual test scripts can be given to the development staff to provide a clear description of the way the application should flow across use cases.

In summary, manual testing fills a gap in the testing repertoire and adds invaluably to the software development process. Types of manual testing techniques are discussed in the following section.

### 3.4.1 Walkthrough

Walkthroughs are similar peer review processes that involve the author of the program, the tester, a secretary and a moderator. The participants of a walkthrough create a small number of manual tracing test cases by simulating a computer. Its objective is to question the logic and basic assumptions behind the source code. Walkthrough is an informal review in which the work product's author describes it to some colleagues and solicits comments.

It is a static analysis technique in which a designer or programmer leads members of the development team and other interested parties through a segment of documentation or code. Participants ask questions and make comments about possible errors, violation of development standards, and other problems. The main thrust of the walkthrough is to detect and document faults; it is not a performance appraisal of the developer. This point must be made to everyone involved so that full cooperation is achieved in discovering errors. The number of errors in production systems decreases by as much as 90 percent in organizations that use walkthroughs diligently. For a junior programmer, working for one year as part of a team that practices walkthroughs may be equivalent to two years of working alone. One of the objectives of walkthrough is to find errors. Other objectives of walkthrough are listed as follows:

- To uncovers errors in function, logic, or implementation for any representation of the software.
- To detect errors early in the software development life cycle.
- To ensure that established procedures have been followed.

- To exchange technical information among project teams which participate in the walkthrough and train them.
- To increase the quality of the project, thereby improving morale of the team members.
- To look for weaknesses or errors of style .
- To verify that the software meets its requirements.
- To ensure that the software has been represented according to predefined standards.
- Make projects more manageable.

A team of at least two members shall be assembled to consider a review as a systematic walkthrough. Roles may be shared among the team members. The participants in walkthrough team are assumed to perform one or more designated tasks of one or more of the following:

- Leader
- Recorder
- Author
- Team member

The walkthrough leader or the author may serve as the recorder. The walkthrough leader may be the author. Individuals holding management positions over any member of the walkthrough team shall not participate in the walkthrough. Walkthrough process is shown in figure 3.4.1. A statement of objectives for walkthrough should be given as an input to the process. Other inputs are the software product being examined and the standards that are in effect for the acquisition, supply, development, operation, and/or maintenance of the software product.

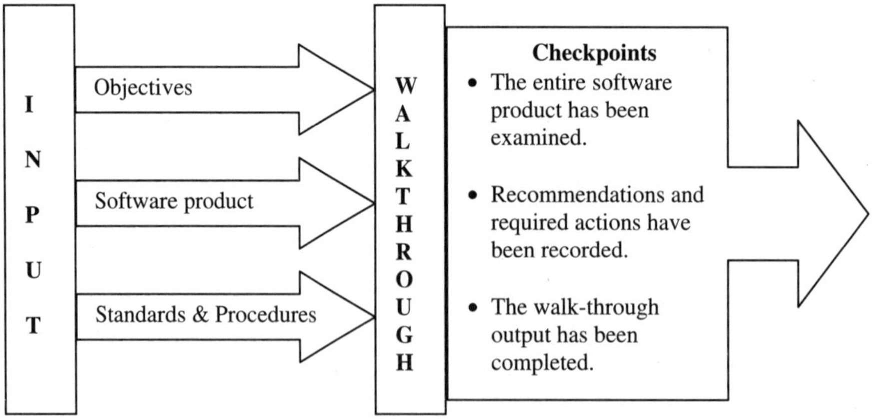


**Figure 3.4.1: Walkthrough Process**

Input to the walkthrough may also include any regulations, standards, guidelines, plans, and procedures against which the software product is to be inspected. The walkthrough shall be considered complete when the entire software product has been examined, recommendations and required actions have been recorded and when the walkthrough output has been completed. A walkthrough differs significantly from inspections because the author takes the dominant role; other specific review roles are usually not defined. Walkthroughs are informal because they typically do not follow a defined procedure, do not specify exit criteria, require no management reporting, and generate no metrics. Table 3.4.1 shows the difference between Walkthrough and Inspection.

**Table 3.4.1: Walkthrough Vs Inspection**

| Inspection | Walkthrough |
|---|---|
| • It is a five-step process.<br>• It is a well-formalized process.<br>• It uses the checklist approach for uncovering errors.<br>• This process takes much longer.<br>• It is a process management tool.<br>• It is used to improve the quality of development process.<br>• It is used to collect defect data to analyze the quality of the process.<br>• During Inspection, written output is important.<br>• Its major role is training the junior staff and transferring expertise. | • It has fewer steps than inspection.<br>• It is less formal process.<br>• It does not use a checklist to guide or any written report to document the team's work.<br>• It does not take longer time.<br>• It is a developer technique.<br>• It is used by development team to improve the quality of product.<br>• It focuses on finding defects. |

Different types of the walkthroughs include specification walkthroughs, design walkthroughs, code walkthroughs, and test walkthroughs. The detailed description of each type is given in the following section.

- ***Specification Walkthrough:*** Specification walkthrough includes system specification, project planning, and requirements analysis. Objective of the specification walkthrough is to check the system specification for problems, inaccuracies, ambiguities and omissions. Participants of specification walkthrough are user, senior analyst, and project analyst and objects used are DFDs, Data Dictionary, ERDs, etc.
- ***Design Walkthrough:*** Design walkthroughs includes preliminary design and detailed design. Objective of the design walkthrough is to check the architecture of the design for flaws, weaknesses, errors, and for omissions. The participants in design walkthrough are User, Analyst, senior designer, Project designers and the objects used for design walkthrough are structure charts, detailed design documents, etc.
- ***Code Walkthrough:*** The objective of Code Walkthrough is to check the code for errors, standards violations, lack of clarity, and for inconsistency. The participants in code walkthrough are author, project programmers, designer, and outside programmers. Objects used in code walkthrough are code listing, compiler listings etc. Code walkthroughs are a sort of guided tour of the program which is led by a

tour guide who describes the points of interest, signposts, family stories, anecdotes, etc. as a walk is being made through the paths of the code.

Code walkthrough may be static as well as dynamic one. In static code walkthrough, documented as an annotated version of the source code with the tour guide comments are shown alongside the code itself. While in dynamic code walkthrough, walkthroughs are conducted by the code reviewers in the meeting as a sort of exploration of the code workings. Both these kinds of walkthroughs have been found to be very effective in training, to introduce new programmers to the code.

- ***Test Walkthroughs:*** Test walkthroughs includes test plan and test procedure. The objective of test walkthrough is to check the testing documents for inadequacy, incompleteness and lack of clarity. The participants in test walkthrough are project programmers, tester, analyst, and designer. Objects used for test walkthrough are test plan, test procedures, sample test data etc.

There are several advantages of Walkthrough, reported variously in the literature. A selection from them is described as follows:

- Walkthrough helps in consciousness-raising as senior people get new ideas and insights from junior people.
- It enables observation of different approaches to software analysis, design, and implementation.
- Walkthrough plays an important role in improving software quality through peer pressure.
- Walkthrough promotes backup and continuity.
- It reduces risk of discontinuity & useless code since several people become familiar with parts of software that they may not have otherwise seen.

In spite of several advantages, many disadvantages of using Walkthrough have been identified including the following:

- Walkthroughs is responsible only for error **detection**, not error correction.
- After the walkthrough **everyone** is responsible for any bugs remaining.
- Walkthrough takes care of the **product**, not the person, is being reviewed.

### 3.4.2 Formal Reviews

Formal reviews are conducted at the end of each phase of software development life cycle. They may also be conducted when a serious problem or concern arises. Two types of formal reviews are available: management reviews and technical reviews. Fundamental differences are listed in Table 3.4.2(a). A formal technical review is a method used to identify defects in an artifact before progressing to the next stage of the development. Unlike system testing, formal technical reviews can take place at any stage in the development life-cycle, and as such allow for the identification, documentation and correction of defects earlier rather than later, therefore saving the company both time and money. Generally, organizations are not inclined to implement formal technical reviews for various reasons. It could be because of the ignorance of the benefits they provide through to the effects they may have on time management aspects of the project. However, standards organizations such as ISO and CMM are pushing companies in the direction of formal technical reviews by insisting on some form of inspection to be in place before offering certification.

Formal technical review methods began in 1976 when Fagan's code inspection technique was published. Fagan inspections consist of 4 pre-defined roles, moderator, scribe, inspector and author. Given the opportunity to follow the basic steps - overview, preparation, inspection, rework and follow up - this method allows for a controlled and documented review, with findings easily referred to and checked upon. This method was considered a major step forward in the field of software quality, and provided IBM with substantial cost savings during the development process. As a result of this, formal technical reviews became popular, and a wide variety of alternatives have since been evolved although often a link to Fagan's code inspection can be seen.

**Table 3.4.2(a): Management Review vs. Formal Review**

| | **Management Review** | **Formal Technical Review** |
|---|---|---|
| **Goal** | To evaluate a project plan or project status relative to that plan. | To examine the product and the result of any assurance activities already conducted on the product. |
| **Purpose** | To ensure the adequacy and completeness of each planning document for meeting project requirements.<br>To ensure that project activities are progressing according to the planning documents, identify the need for corrective action to the plan or the project, and ensure proper allocation of resources. | To evaluate the software elements such as requirements and design.<br>To ensure conformity to its specifications, compliance of the development of the software elements with its plans, and the integrity of changes to the software elements. |
| **Result** | Results of the management reviews are summarized in a management review report and are auditable and traceable to and from the appropriate planning documents. | Results of the technical reviews are summarized in technical review reports, which are auditable and traceable to and from the appropriate planning documents. |

The objective of the technical review is to search out design weaknesses, faulty designs, or designs which may be cost drivers. Problems can often be identified early from the data item submissions, and attempts can be made to clarify these problems on a case-by-case basis before the review to avoid major problems and surprises at the meetings. The necessity for a technical or progress review is determined in part by the type and characteristics of the system being built. A suggestive guideline includes following:

- Review the product, not the producer.

- Keep on track, don't drift.
- Limit debate and rebuttal.
- Enunciate problem areas, but don't attempt to solve the problems.
- Take written notes.
- Insist upon advance preparation. If your reviewers aren't prepared, cancel the meeting and reschedule.
- Consider developing a checklist for the product being reviewed.
- Determining whether or not the software element corresponds to the specifications and standards to which it must adhere, and recording any deviations
- Listing issues, recommendations, and responsible individuals for resolving the issues
- Identifying other issues that need to be addressed
- Documenting the meeting, deficiencies found in the software element, and recommendations for management

However difficult to implement it may be, it has several advantages to count and benefit from. Selections of some major advantages are listed as follows:

- Formal technical review provides opportunity to change course of a project before start of next phase.
- Since scope of review is usually broad, it gives opportunity to recognize global problems.

In spite of several advantages, there are also disadvantages of using Formal Technical Review. Some of the disadvantages are listed as follows:

- If participants do not have materials ahead of time and spend time preparing, Formal Technical Review will accomplish little or nothing.
- Formal Technical Review focuses on major issues and miss the minor technical issues.

**Formal Technical Reviews (FTR)** are performed in typical stages. Figure 3.4.2 depicts the process for technical review that involves these typical reviews, listed as follows:

- System Requirements Review (SRR)
- Software Specification Review (SSR)
- Preliminary Design Review (PDR)
- Critical Design Review(CDR)

**Technical Review Checklist:** The purpose of technical reviews is to analyze and evaluate the software artifacts such as requirements and design to ensure conformity to its specifications, compliance of the development of the software elements with its plans, and the integrity of changes to the software elements. It is required to verify whether or not activities were performed in accordance with their respective plans. In order to accomplish the desired need, a checklist may be provided to review both the software product and plans for assurance activities. Table 3.4.2(b-e) depicts the technical review checklists.

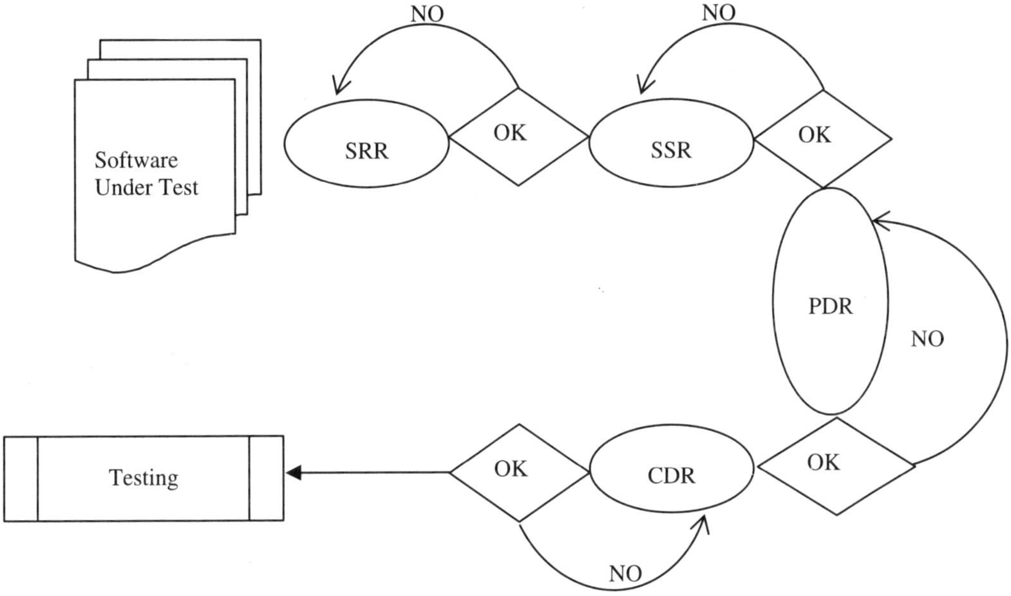


**Figure 3.4.2: Technical Review Process**

**Table 3.4.2(b): Software Requirement Review Checklist**

| S.N | Checks | Yes | No |
|---|---|---|---|
| 1 | Whether the functional requirements cover all abnormal situations? | | |
| 2 | Whether the compatibility of models, algorithms, and numerical techniques has been checked? | | |
| 3 | Whether the adequacy of requirements for man-machine interface has been checked? | | |
| 4 | Whether the design/implementation constraints have been justified? | | |
| 5 | Whether the feasibility of design, operation, and maintenance of software has been checked? | | |
| 6 | Whether the unique requirements have been defined more than once? | | |
| 7 | Whether the requirements for fault tolerance and graceful degradation have been identified? | | |
| 8 | Whether the requirement specification is efficient and forms the right basis for detailed design specification and functional test cases? | | |
| 9 | Whether the requirements are verifiable? | | |

**Table 3.4.2(c): Software Design Review Checklist**

| S.N | Checks | Yes | No |
|---|---|---|---|
| 1 | Whether the design features are consistent with the requirements? | | |
| 2 | Whether the design implements required program behavior with respect to each program interface? | | |
| 3 | Whether the required processing steps are included? | | |
| 4 | Whether the possible outcomes of each decision point have been designated? | | |
| 5 | Whether all expected situations and conditions have been taken into account during the design? | | |
| 6 | Whether the integrity of changes has been ensured for design configuration? | | |
| 7 | Whether each program has a single function? | | |
| 8 | Whether the design is structured? | | |
| 9 | Whether unnecessary design complexity and representation has been avoided? | | |

**Table 3.4.2(d): Source Code Review Checklist**

| S.N | Checks | Yes | No |
|---|---|---|---|
| 1 | Whether 'complete and precise code implementation of design' is viable? | | |
| 2 | Whether code is integrated and debugged to verify design specification? | | |
| 3 | Whether the code creates the required database? | | |
| 4 | Whether the un-referenced or undefined variables, constraints or data types have been identified? | | |
| 5 | Whether the code is logically consistent with design? | | |
| 6 | Whether the code conformance to specified standards has been checked? | | |
| 7 | Whether the variables are properly specified and used? | | |
| 8 | Whether the accuracy of comments has been checked? | | |
| 9 | Whether the programs invoked with the correct number of parameters? | | |
| 10 | Whether the cross-references or data dictionaries are included showing variables and constant access by the program? | | |
| 11 | Whether the code consists of program with one entry point and one exit point? | | |
| 12 | Whether the syntax and semantics of the language has been checked in which the code has been written? | | |
| 13 | Whether the presence of unintended infinite loop has been checked? | | |
| 14 | Whether the unwanted recursion has been eliminated form the code? | | |
| 15 | Whether the code protects against detectable run-time errors? | | |
| 16 | Whether the comment statements adequately describe each routine? | | |

**Table 3.4.2(e): Test Readiness Review Checklist**

| S.N. | Checks | Yes | No |
|---|---|---|---|
| 1 | Whether the recommended changes have been made to the code. | | |
| 2 | Whether the error rate has been checked to begin for the next type of testing? | | |
| 3 | Whether the test cases and procedures are checked and complete? | | |
| 4 | Whether the facility for test has been checked? | | |
| 5 | Whether the test tools have been checked? | | |
| 6 | Whether the test procedures have been checked? | | |

### 3.4.3 Inspection

Software inspection is one of the most important manual techniques which allow testing the program without running and comparing code or design of work product to set of pre-established inspection rules. A software inspection is a expert-group review process that is used to detect and correct defects in a software work product. It is a formal-technical activity performed by the work product author and a small peer group on a limited amount of material. Most of the testing techniques available are intimately related to the system attribute whose value they are designed to measure. Therefore, they generally do not offer any information about other attributes. Some software engineers report inspection as adequate for any kind of document, for example, specifications, test plans, etc.

The subject of the inspection is typically a document such as a requirements specification or a test plan, and the purpose is to find problems and see what's missing, not to fix anything. Attendees should prepare for this type of meeting by reading the document; most problems found during its preparation. Thorough preparation for inspections is difficult, painstaking work, but is one of the most cost effective methods of ensuring quality. Inspection has been identified as one of the top nine software industry best practices. Nearly 30 years of experience shows that inspections are efficient and effective ways to discover defects in any work product.

A team that examines the product of a particular simulation development phase normally performs inspection. Inspection team normally consists of four or five members, including a moderator or leader, a recorder, a reader who is the representative of the Developer and who presents the material being inspected, the verification and validation Agent, and one or more appropriate subject matter experts.

**Inspection Process:** Figure 3.4.3(a) shows the inspection process where the inputs to the inspection are as follows:

- A statement of objectives for the inspection.
- The software product to be inspected.
- Documented inspection standards and procedures.

- Inspection reporting forms.
- Current anomalies or issues list.

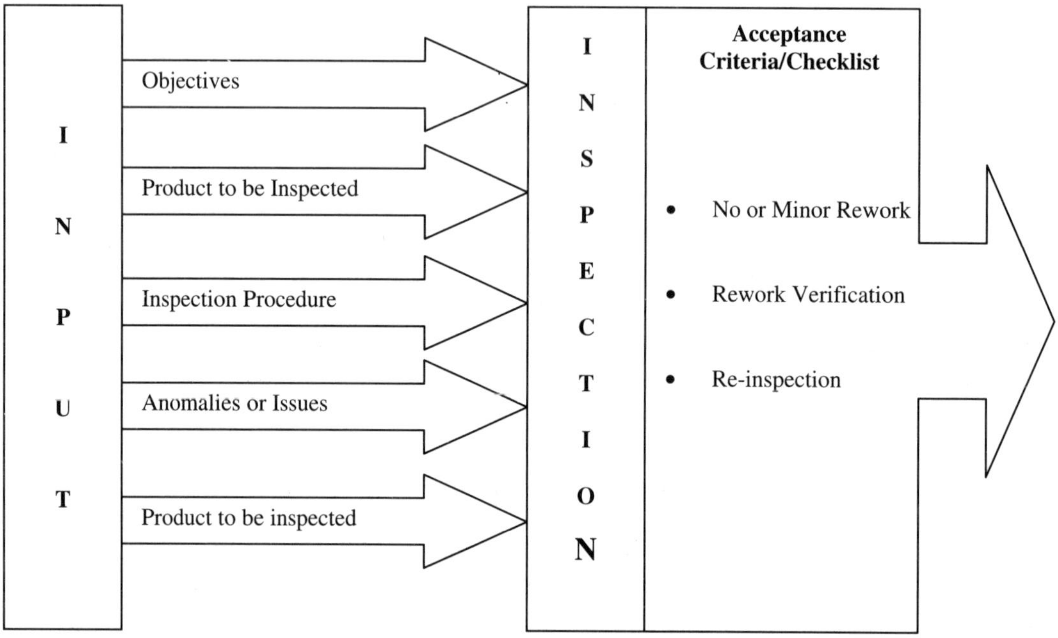


**Figure 3.4.3(a): Inspection Process**

In spite of these essential inputs to the inspection, some of the other desired inputs may also be required for the inspection process. Such inputs may possibly include following:

- Inspection checklists
- Regulations, standards, guidelines, plans, and procedures against which the software product is to be inspected
- Hardware product specifications
- Hardware performance data

The exit decision shall determine if the software product meets the inspection exit criteria and shall prescribe any appropriate rework and verification. Inspection team verifies the inspected software product against the following checklist:

- *Accept with no or minor rework.* The software product is accepted as is or with only minor rework that would require no further verification.
- *Accept with rework verification.* The software product is to be accepted after the inspection leader or a designated member of the inspection team other than the author verifies rework.

- *Re-inspect.* Schedule a re-inspection to verify rework. At a minimum, a re-inspection shall examine the software product areas changed to resolve anomalies identified in the last inspection, as well as side effects of those changes.

**Inspection Phases:** Normally, an inspection consists of five phases: overview, preparation, inspection, rework, and follow-up. Figure 3.4.3(b) depicts the process including these phases.

**Inspection Team:** Software inspection is a group activity and is being carried out by inspection team. Software inspection team is made up of at least four members including:

- *Author:* A key person who is the author of the code being inspected.
- *Inspection Leader:* An active member of the team who finds errors, omissions and inconsistencies.
- *Reader:* Most responsible member who reads the code to the team.
- *Moderator:* A person who chairs the meeting and preparing notes on discovered errors.

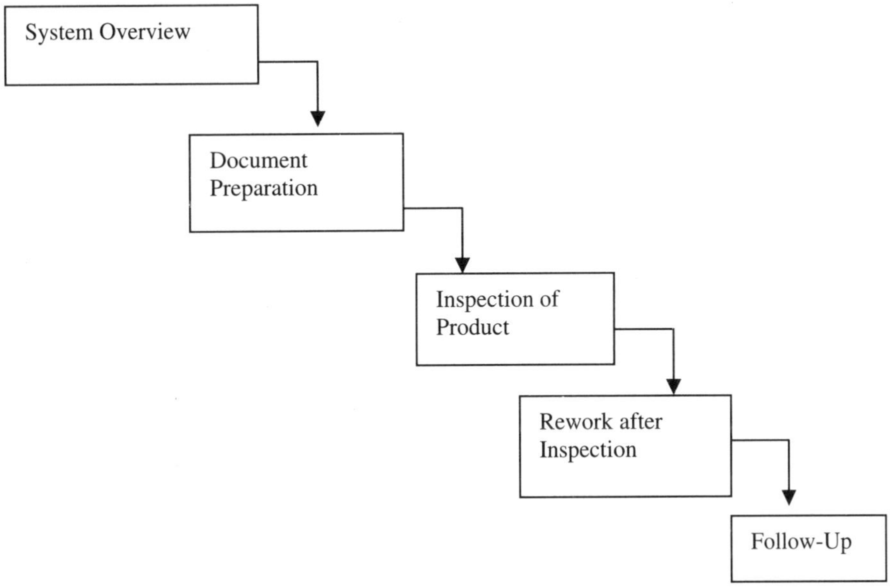


**Figure 3.4.3(b): Inspection Phases**

All participants in the review are inspectors but author shall not act as inspection leader and should not act as reader or recorder. Individuals holding management positions over any member of the inspection team shall not participate in the inspection. The individuals may make additional reference material available responsible for the software product when requested by the inspection leader. There are several types of inspection reported in the literature including the following:

- Code Inspection;
- Design Inspection;
- Architectural Inspections; and
- Test Inspections

**Methodology:** Formal Inspection of system development products is a very effective means of reducing errors in software development. The ultimate goal of the inspection is to remove errors as near source as possible hence reducing costs of rework. Software inspection involves people examining the source representation with the aim of discovering anomalies and defects. It is a very effective technique for discovering errors. Inspection does not require execution of a system, and hence may be used before implementation. Software inspection may be applied to any representation of the system (requirements, design, test data, etc.). It is a formalised approach to document reviews, and is intended explicitly for defect detection not correction. These defects may be logical errors, anomalies in the code that might indicate an erroneous condition (for emaple. an uninitialized variable) or non-compliance with standards. A set of activities to be performed during the inspection process include following:

- An exact system overview is presented to the Inspection team members.
- Readers paraphrases are designed and described how to implement.
- Source code and related documents are distributed to the members of the inspection team in advance.
- Discovered errors during the inspection are precisely noted for further reference.
- Questions raised during discourse are pursued until error are recognized.
- Discovered errors are referred and modifications made to uncover and repair the discovered errors.
- As per the result, re-inspection may or may not be required.

Following the activities as per prescription for the inspection process, a range of decisive actions are possible. A selection of the most probable actions include following:

**Re-inspection:** Inspect modified products to avoid "bad fix" problem.

**Inspection Scheduling:** Inspections should be scheduled with full care.

**Do not Postpone:** The result of postponing inspection is usually lengthening of the overall schedule and hence may increase overall product cost.

**Sessions:** Two sessions of two hours in a day is recommended.

**Advantages:** Numerous advantages of Inspection have been reported in the literature. Some pertinent once are listed as follows:

- Many different defects may be discovered in a single inspection. While in testing, one defect, may mask another so several executions are required.
- Software inspection reuse domain and programming knowledge so reviewers are likely to have seen the types of error that commonly arise.

- It provides comprehensive statistics on classes of errors.
- It Increases product quality.
- It is highly effective for projects of all sizes.

**Disadvantage:** In spite of various advantages, following disadvantages of Inspection have been broadly identified:

- Inspections can check conformance with a specification but not conformance with the customer's real requirements.
- Inspections cannot check non-functional characteristics such as performance, usability, etc.
- Inspectors must be independent of programmers.
- Programmers may feel inspection is a personal attack on their work.
- It is time consuming task.

## 3.5 AUTOMATED TECHNIQUES

Automatic testing is used to replace or supplement manual testing techniques with a suite of test tools. Benefits of using automatic testing include increased software quality, improved time to market, repeatable test procedures, and reduced testing costs. A typical automated test suite may run to completion fast, without any human intervention required. For a sophisticated product, manual testing may require dozens of staff months to perform for the similar level of testing. With a complex testing process manual testing may yield inconsistent coverage and results depending on the staff and schedule employed. On the other hand, an automated test suite ensures the same scope and process is used repeatedly each time testing is performed.

Automated testing has an upfront cost to develop. An average automated test suite development is normally 3-5 times the cost of a complete manual test cycle. Over multiple product releases with multiple cycles per release, this cost is quickly recouped. With its much shorter execution time an automated test suite can be run multiple times over the course of a product development cycle. By testing earlier and more often bugs are detected and corrected earlier and at much reduced expense. The sum of improved test procedures and testing productivity is a substantial improvement in product quality. Automated testing detects functional and performance issues more efficiently, allowing test staff to focus on quality in areas such as documentation, installation, hardware compatibility, etc. Following section describes the types of automated testing techniques. A variety of techniques are employed for automated static testing, including ones described briefly in following sections.

### 3.5.1 Syntax Parser

Atomic static analysis tools may be used for parsing the syntax through splitting the program/documents text into individual statements. Syntax parser evaluates the consistency of the statement by checking the program or documents text internally. The selection of test cases in syntax testing could start with single-error sentences. This is likely to reveal most faults assuming the faults are mutually independent and a fault is triggered by one error in a sentence. After all the sentences with one error are tried, the testing proceeds to pairs of errors, three error combinations, and so on. The number of test cases grows exponentially by the number of combined errors. Different kinds of errors can be produced as a result of syntax parsing including following:

- **Syntax Errors:** Syntax errors may exist at the top-level, intermediate-level or field-level in the grammar hierarchy. It violates the grammar of the underlying language. They are created by removing an element, adding an extra element and providing the elements in wrong order.
- **Delimiter Errors:** Delimiters mark the separation of fields in a sentence. In ASCII-coded languages the fields are normally characters and letters, and delimiter are *white space* characters. Delimiters can be omitted, multiplied or replaced by other unusual characters. Paired delimiters, such as braces, can be left unbalanced.
- **Field Value Errors:** An illegal field in a sentence is called a field-value error. Normally, it has a *range* or many disjoint ranges of allowable values, and include values, which are one-below, one-above and totally out-of-range. Values exactly at the range boundary should also be checked.
- **Context Dependent Errors:** A context-dependent error violates some property of a sentence, which cannot, in practice, be described by context-free grammar.
- **State Dependency Error:** A state dependency error is generated by inputting a correct sentence during an incorrect state.

### 3.5.2 Static Verification

Static verification technique aims at detecting problems in the translation between specification and program realization. This technique requires formal specifications and formal definitions of the specification and programming languages used. Static verification techniques involve examination and analysis of the program for error detection. Static verification technique does not provide any means to check whether the program actually solves the given problem, and whether the given specifications are correct or not. The actual values provided by the program are compared with the target values as pre-defined in the specification document during static verification technique. Since static verification is based on formal methods, it leads to objective and correct result. However, it is very difficult and time-consuming to elaborate the formal specifications for static verification. Therefore, static verification is being used for software that needs to be highly reliable.

### 3.5.3 Symbolic Execution

Symbolic execution is another important technique which is normally subsumed under static analysis and is most appropriate for the analysis of mathematical algorithms. The whole classes of values can be represented by a single interpretation by making use of symbolic values only. This leads to a very high coverage of test cases. Symbolic execution analyses what a program does along a given path by computing the values of a program's variables as functions which represent the sequence of operations carried out as execution. Symbolic execution technique is mainly used for testing numerical programs as it is very expensive.

## 3.6 STATIC Vs DYNAMIC TESTING

Static analysis is usually faster than dynamic analysis but less precise. Therefore it is often desirable to retain information from static analysis for run-time verification, or to compare the results of both the techniques. However, this requires writing two programs, which may not act identically under the same conditions. It would be desirable to share the same generic algorithm by static and dynamic testing. One of the major problems that most companies experience with static analysis is the enormous amount of anomalies that are generated by the analyzer. Some companies discard the use of static analysis just by looking at the size of their first report. Many anomalies can be discarded at a glance if coding conventions are introduced. The static analyzers have problems recognizing certain code constructs like function pointers, and identifying variables in condition statements. Furthermore, the analyzers cannot see what goes on outside the unit being analyzed in the operating system, and libraries. By giving easily recognizable names or suffixes to variables of this nature, anomalies that are reported can be easily spotted and discarded.

Dynamic testing is much more appropriate for run-time analysis as it looks at the effects of the program's input parameters and investigates its actual run-time behavior. A complete testing tool would execute the test program for every single set of input parameters and measure its execution time. Dynamic testing differs from static testing in number of ways. Static tools offer greater marginal benefits. Static testing beats dynamic testing by a wide margin. Static testing gives comprehensive diagnostics for code and finds all of the followings while dynamic testing may not do so:

- Syntax error
- Code that is hard to maintain
- Code that is hard to test
- Code that dose not conform to coding standards
- ANSI violation

For dynamic testing the test objects are executed or simulated. Dynamic testing is an imperative process in the software life cycle. Every procedure, every module and class, every subsystem and the overall system must be tested dynamically, even if static tests and program verifications have been carried out. The activities for dynamic testing include the following:

- Preparation of the test object for error localization
- Availability of a test environment
- Selection of appropriate test cases and data
- Test execution and evaluation

Static analysis followed by static testing approximates the set of possible program states and includes abstract interpretation, which typically encompasses a wider range of behaviors than the original, using an abstract version of the program. This analysis iterates over the abstract states until a fix point is reached or a certain limit, such as a second loop execution, is reached.

Static testing differs from dynamic testing, which evaluates properties against an event trace originating from a concrete program execution. Dynamic testing techniques actually execute the system under test. This has the key advantage of having precise information available. Some fully automated dynamic analysis algorithms even only require a single execution trace to deduce possible errors. If a test suite with high coverage exists, run-time verification can thus explore a large part of the possible behaviors while scaling significantly better than software model checking. Dynamic analysis requires an execution environment. Table 3.6 lists the differences between static and dynamic testing.

**Table 3.6: Static Testing Vs Dynamic Testing**

| Static Testing | Dynamic Testing |
|---|---|
| Static testing is about prevention. | Dynamic Testing is about cure. |
| Static testing may prove to be more cost effective. | Dynamic testing is less cost effective. |
| Static testing may achieve 100% statement coverage in relatively short time. | Dynamic testing often achieves less than 50% statement coverage because dynamic testing finds bugs only in part of codes that are actually executed. |
| Static testing usually takes shorter time. | Dynamic testing may involve running several test cases, each of which may take longer than compilation. |
| Static testing may uncover variety of bugs. | Dynamic testing uncovers limited type of bugs. that are explorable through execution. |
| Static testing can be done before compilation. | Dynamic testing can take place only after executables are ready. |

## 3.7 CONCLUSION

Static testing of programs is a proven technology in the implementation of compilers and interpreters. Recent years have begun to see application of static testing techniques in areas such as software validation and software re-engineering. In dynamic testing code is required at the time of testing. Therefore analysis will be performed on executing programs, and hence cannot be done in early stage. On the other hand, static analysis can be applied early in the software development life cycle. Static analysers are software tools for source text processing. Static testing is done for the purpose of validating that a work product such as a design specification properly implements all the system requirements, and verifying the quality of the design. Static testing is increasingly recognized as a fundamental tool for high performance implementations and verification of programming languages and systems static testing is the term applied to the testing of computer software that is performed without actually executing programs built from that software.

In most cases the testing is performed on some version of the source code and in the other cases some form of the object code. The term is usually applied to the testing performed by an automated tool, with human testing being called program understanding or program comprehension. Essential functions of static testing include checking whether representations and descriptions of software are consistent, noncontradictory and unambiguous. Static testing generally takes care of correct descriptions, specifications and representations of software systems. Therefore, static testing is said to be a precondition to any further testing exercise. Walkthrough is a static testing technique in which the work product's author performs informal review and describes it to some colleagues and solicits comments. A Walkthroughs differ significantly from inspections because the author takes the dominant role. Walkthroughs are informal because they typically do not follow a defined procedure, do not specify exit criteria, require no management reporting, and generate no metrics. While software inspection is a static testing technique that relies on visual examination of development products to detect errors, violations of development standards, and other problems.

## KEY TERMS

Static testing, Verification, Validation, Static analyzer, Strategy, Static analysis technique, Inspection, Walkthrough, Static verification, Syntax parser, Symbolic Execution, Principles of static testing, Control flow analysis, Data flow analysis, Interface analysis, Information flow analysis, Path analysis, Complexity, Specifications, Design walkthrough, Code walkthrough, Test walkthrough, Formal review, Management review, Static testing, Dynamic testing.

## REVIEW QUESTIONS

## OBJECTIVE TYPE QUESTIONS

1. Static testing is done for_______________.
   - (a) Validation purpose only
   - (b) Verification purpose only
   - (c) Verification and validation purpose both
   - (d) None of the above
2. Static testing involves_______________.
   - (a) Inspection and walkthrough
   - (b) Walkthrough and peer-review

(c) Inspection, Walkthrough, and peer-review

(d) None of the above

3. For the languages that have strong type checking, static testing is_______________.

(a) Less cost effective

(b) High cost-effective

(c) Not cost-effective

(d) None of the above

4. A peer-review process that involves the author of the program, the tester, a secretary and a moderator is_______________.

(a) Inspection process

(b) Walkthrough process

(c) Syntax parser

(d) Static verification process

5. Inspection is a formal technical activity that is performed to_______________.

(a) Detect the errors only

(b) Correct the errors only

(c) Detect and correct the errors

(d) None of the above

6. For testing numerical programs, the technique used is called_______________.

(a) Walkthrough

(b) Inspection

(c) Symbolic Execution

(d) Syntax parsing

7. _______________ uses the checklist approach for uncovering errors.

(a) Inspection

(b) Walkthrough

(c) Both Inspection and Walkthrough

(d) None of the above

8. Code walkthrough may be_______________.

(a) Static only

(b) Dynamic only

(c) Both static and dynamic

(d) None of the above

9. Design walkthrough includes_______________.
    (a) Preliminary design
    (b) Detailed design
    (c) Both preliminary design and detailed design
    (d) None of the above
10. Test walkthrough includes_______________.
    (a) Test plan only
    (b) Test procedures only
    (c) Both test plan and procedures
    (d) None of the above
11. Walkthrough is responsible for_______________.
    (a) Error detection only
    (b) Error correction only
    (c) Both error detection and correction
    (d) None of the above
12. Formal reviews are conducted at_______________.
    (a) The beginning of each phase of development life cycle
    (b) The end of each phase of development life cycle.
    (c) The beginning of the testing phase of development life cycle.
    (d) None of the above

## SHORT ANSWER TYPE QUESTIONS

1. What do you mean by static testing of the program?
2. Distinguish between static testing and dynamic testing.
3. How do static analyzer work?
4. How can static testing save substantial time and cost of the development?
5. What do you mean by strategy in general?
6. Write the need, importance and significance of software testing strategies.
7. Propose a strategic approach to software testing.
8. Identify the factors related to verification and validation process.
9. How does static testing analyzes the unreachable code and unconditional branches into loop?
10. What are the techniques available for static testing? Discuss any two.

11. Static testing is the pre-condition to any further testing exercises. Justify the statement.
12. What do to mean by software inspection?
13. Write the major advantages and disadvantages of software inspection.
14. How can you perform walkthrough? List three advantages and disadvantages each of walkthrough.
15. How can syntax parser evaluate the consistency of the statement?
16. What are the different static verification techniques available?
17. Elaborate the formal specification for static verification.
18. What does symbolic execution do?
19. Write down the principles of static testing.
20. List few testing metrics extracted form code without dynamic execution of source code.
21. How can you evaluate robustness and complexity of the source code?
22. How does violations of set coding standards affect overall testing process?
23. How can unreachable code, unreferenced procedures and unused variables be removed?
24. What are data flow anomalies? List five such anomalies.
25. How can complexity of code be reviewed and evaluated?
26. How can walkthroughs help in detecting errors early in the software development life cycle?
27. How can walkthrough process increase the quality of software?
28. Distinguish between inspection and walkthrough with respect to effort, time and resources.
29. What is specification walkthrough? Explain with an example.
30. How can walkthroughs help in consciences raising?
31. When do we perform formal reviews generally?
32. Distinguish between system testing and formal technical review.
33. Distinguish between managerial review and formal technical review.
34. List two anomalies or issues with respect to inspection.
35. What do you mean by re-inspection?

## DESCRIPTIVE TYPE QUESTIONS

1. Static testing is less cost-effective for languages that have strong type checking like java. Comment.

2. How can static testing serve the purpose with respect to validation and verification?
3. Discuss various techniques available for static testing. Suggest the most appropriate technique.
4. Though, static verification technique does not provide any means to check whether the program actually solves the given problem, it is the most demanding technique. Justify.
5. Static verification can be used for verifying reliability of the software. Comment.
6. Propose a set of strategies for static testing not included in this chapter with proper justification.
7. Describe walkthrough process elaborating each step in detail. Take any small project, perform walkthrough and show the outcome.
8. Why do formal technical review only identify the problems, but do not attempt to solve it?
9. Prepare a checklist for the product under review process.
10. Propose a checklist for the product under inspection.

## REFERENCES

1. M.E. Fagan "Design and Code Inspections to Reduce Errors in Program Development", IBM Systems Journal, March 1976.
2. A.F. Ackerman, L. Buchwald, and F. Lewski, "Software Inspections: An Effective Verification Process," IEEE Software, May 1989.
3. Marilyn Bush, "Improving Software Quality: The Use of Formal Inspections at the Jet Propulsion Laboratory", Proceedings of the 12th
4. International Conference on Software Engineering, pages 196-199, IEEE Computer Society Press, Nice, France, March 1990.
5. Curtis, Bill, Herb Krasner, Vincent Shen, and Neil Iscoe. (1987). "On building software process models under the lamppost." Proceedings of the 9th International Conference on Software Engineering (pp. 96–103). Monterey, CA: IEEE Computer Press Society.
6. Lewis, William E. (2000). Software Testing and Continuous Quality Improvement. Boca Raton, FL: Auerbach.
7. McConnell, Steve. (1996). Rapid Development: Taming Wild Software Schedules. Red-mond, WA: Microsoft Press.
8. Pfleeger, Shari Lawrence. (2001). Software Engineering: Theory and Practice (2nd ed.). Upper Saddle River, NJ: Prentice Hall.

9. Schulmeyer, G. Gordon and Garth R. Mackenzie. (2000). Verification and Validation of Modern Software-Intensive Systems. Upper Saddle River, NJ: Prentice Hall.

10. Bhandari, M. J. Halliday, J. Chaar, K. Jones, J. Atkinson, C. Lepori-Costello, P. Y. Jasper,

11. E. D. Tarver, C. C. Lewis, and M. Yonezawa. In-process improvement through defect data interpretation. The IBM System Journal, 33(1):182–214, 1994.

12. B. W. Boehm. Software Engineering Economics. Prentice Hall, Englewood Cliffs, NJ, 1981.

13. J.K. Chaar, M.J. Halliday, I.S. Bhandari, and R. Chillarege. In-process evaluation for software inspection and test. IEEE Transactions on Software Engineering, 19(11):1055–1070, November 1993.

14. D. M. Cohen, S. R. Dalal, M. L. Fredman, and G. C. Patton. The AETG system: An approach to testing based on combinatiorial design. IEEE Transactions on Software Engineering, 23(7):437–444, July 1997.

15. P. G. Frankl and O. Iakounenko. Further empirical studies of test effectiveness. In Proceedings of the ACM SIGSOFT 6th International Symposium on the Foundations of Software Engineering (FSE-98), volume 23, 6 of Software Engineering Notes, pages 153–162, New York, Nov. 3–5 1998. ACM Press.

16. T. Graves, M. J. Harrold, J.-M. Kim, A. Porter, and G. Rothermel. An empirical study of regression test selection techniques. In Proceedings of the 20th International Conference on Software Engineering, pages 188–197. IEEE Computer Society Press, April 1998.

17. P. A. Hausler, R. C. Linger, and C. J. Trammell. Adopting cleanroom software engineering with a phased approach. IBM Systems Journal, March 1994.

18. Independent Assesment Team. Mars program independent assessment team summary report. Technical report, 2000.

19. A. Jaaksi. Assessing software projects: Tools for business owners. In Proceedings of the 9$^{th}$ European Software Engineering Conference held jointly with 10th ACM SIGSOFT International Symposium on Foundations of Software Engineering (ESEC/FSE 2003), pages 15–18. ACM Press, September 2003.

20. J.-M. Kim and A. Porter. A history-based test prioritization technique for regression testing in resource constrained environments. In Proceedings of the International Conference on Software Engineering (ICSE 2002), May 2002.

21. J. L. Lions. Ariane 5, flight 501 failure, report by the inquiry board. Technical report, 1996.

22. B. Marick. The Craft of Software Testing: Subsystems Testing Including Object-Based and Object-Oriented Testing. Prentice-Hall, 1997.

23. J. Nielsen. Designing Web Usability: The Practice of Simplicity. New Riders Publishing, Indianapolis, 2000.

24. T. J. Ostrand and M. J. Balcer. The category-partition method for specifying and generating functional tests. Commun. ACM, 31(6):676–686, June 1988.

## USEFUL LINKS

http://www.csaengineering.com/dyntest/dyntest.shtml

http://www.irisa.fr/manifestations/2006/CSTVA06/index.htm

www.sqe.com/certification.asp?f=dis&ci=stf

www.cigital.com/labs/reliability/testing.php

www.softwareqatest.com/qatweb1.html

portal.acm.org/ citation.cfm?coll=GUIDE&dl=GUIDE&id=1007514

www.sqe.com/getcertified

www.businessanalysisbooks.com/0672319837.html

www.aptest.com/glossary.html

www.cs.pitt.edu/coco/papers/woss-2004.pdf

www.actapress.com/PDFViewer.aspx?paperId=14577

www.artima.com/weblogs/viewpost.jsp?thread=7590

sunset.usc.edu/~rroshand/papers/CDSA01.pdf

fmv.jku.at/papers/ArthoBiere-ETH-TR-466-2005.pdf

www.cs.virginia.edu/woda2004/papers/ostrand.pdf

www.fit.ac.jp/~zhao/staplab/afanti-home.html

www.ist.tugraz.at/publications/psfiles/ckpw04.pdf

www.cis.strath.ac.uk/teaching/ug/classes/52.422/static_and_dynamic_analysis.pdf

sunset.usc.edu/classes/cs377_2004/Lecture14.ppt

www.cs.kent.ac.uk/pubs/2000/1277/index.html

pcsostres.ac.upc.edu/micro38/dynamic.php

www.cs.wm.edu/~coppit/csci780-fall2003/ final-papers/hewett-dipalma.pdf

www.laatuk.com/tools/testing_tools.html

leshatton.org/index_PR.html

opensourcetesting.org/performance.php

csdl.computer.org/comp/proceedings/ wcre/2004/2243/00/22430005.pdf

## Chapter 4

# Black Box Testing

*It is not enough to do your best. You must know what to do and then do your best.*
**W. Edwards Deming**

## OBJECTIVES

A successful completion of this chapter will enable the reader for the following:

- To understand the term 'Black-Box'.
- To realize the importance of Black-Box testing.
- To comprehend the role of Black-Box testing technique in software engineering.
- To list the advantages and disadvantages of Black-Box testing technique.
- To decide the order of the tests to be performed.
- To elaborate the scope and prospects of equivalence partitioning.
- To design test cases for equivalence partitioning.
- To visualize the need for equivalence partitioning.
- To understand and establish how to identify equivalence classes.
- To appreciate the goals and benefits of boundary value analysis.
- To define the term boundary conditions.
- To specify the advantages and disadvantages of boundary value analysis.
- To design test cases for boundary value analysis.
- To list the objectives of robustness testing.
- To realize the importance of robustness testing, and its methodology.
- To design the test cases for robustness testing.
- To realize the need of syntax testing.
- To elaborate different types of errors under syntax testing.
- To prepare Black-Box test suite for testing a given software.

## 4.1 INTRODUCTION

Computer technologies play an important role in the modern society. Computers and software, that drives them, affect more people and their businesses today. Therefore, there is a demand not only to build software systems quickly, but to focus on quality issues too. Low quality software that can cause loss of life or money is no longer acceptable. Testing has taken an important place in software development practices. As a process by itself it is related to two other processes verification and validation. Verification is the process of evaluating a software system or component to determine whether the products of a given development phase satisfy the conditions imposed at the start of that phase. Validation is a process of evaluation a software system or component during or, at the end of, the development cycle in order to determine whether it satisfies specified requirements.

Software testing is an extensive and difficult process for any realistic software system. It is a process or several processes designed to make sure computer code does what it was designed to do and that it does not do anything unexpected. By automating the test generation process, the overall cost can be significantly reduced. The software testers are responsible to design tests that reveal defects and can be used to evaluate usability and reliability of the software performance. To achieve these goals testers must select a finite number of test cases. There are several techniques to design test cases including the following two:

- White-Box which is also called clear or glass box technique.
- Black-Box, which is also called functional, or specification technique.

**Figure 4.1(a): A Simple Black Box Specification**

The White-Box technique focuses on the internal structure of the software under test. The tester, while using this technique, should know the internal structure of the software under test. The source code or pseudo code must be available. On the other hand, in black-box approach, the software is viewed as a Black Box, as shown in the figure 4.1(a). The goal of a tester is to be completely unconcerned about internal structure of the software. Instead, testers concentrate on software behavior and functionality. Table 4.1 symbolizes the knowledge source for Black-Box technique.

Black box testing begins with a metaphor. Imagine an electronics system housed in a Black Box with lights, switches, and dials on the outside. This electronic system must test without opening it up. It can be verified if it works just by flipping switches (inputs) and what happens to the lights and dials (outputs). This is nothing but the testing. Black Box software testing does the same thing with software. Black Box testing of software is testing without knowledge of the internal workings of the software being tested. Whenever a Black Box testing is applied to software engineering, the tester would only know the 'legal' inputs and what the expected outputs should be, but not how the program actually arrives at those outputs. Therefore, the tester and the programmer can be independent of one another in Black-Box testing. This will avoid programmer's biasness toward his own work.

**Table 4.1: Black-Box Technique and Knowledge Source**

| **Technique** | **Knowledge Source** |
|---|---|
| Black-Box | • Requirement Document<br>• Specifications<br>• Domain Knowledge<br>• Defect Analysis Data |

Black-Box testing can be considered testing with respect to the specifications and no other knowledge of the program is necessary for this kind of the testing. Selection of test data as well as the interpretation of test results is performed on the basis of the functional properties of a piece of software in case of Black Box testing. Black Box testing should not be performed by the author of the program who knows too much about the program internals.

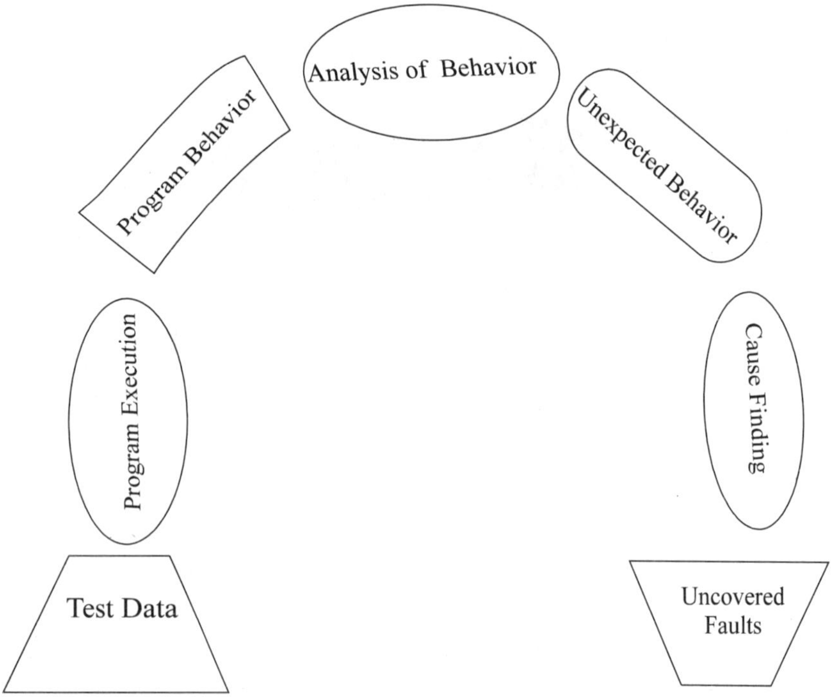


**Figure 4.1(b): Black-Box Testing**

Black-Box or Functional class of testing techniques use the specifications or the required behavior of the software as a starting point to design test cases. Functional testing is the process of verifying that a system or system component adheres to the specification that defines its requirements as shown in the figure 4.1(b). Functional testing can be performed at the system level or the unit level. To perform functional testing, it is required to create a set of input/outcome relationships that verify whether each specification requirement is implemented correctly. At least one test case should be created for each entry in the specification document and should be executed to verify whether the correct outcomes are produced. These test cases should test the various boundary conditions for each entry.

## 4.2 BLACK-BOX TECHNIQUES

Black-Box testing is a way to test software so that the internal workings of the item being tested are not known by the tester and is based solely on the knowledge of the system requirements. It is not a type of testing technique; it instead is a class of testing techniques, which does not need any knowledge of internal design or code. The types of testing under this class are totally based on the testing for requirements and functionality of the work product or software application. The major objectives of the Black Box testing is to uncover the following type of errors:

- Incorrect or missing functions
- Interface errors
- Errors in data structures or external database access
- Performance errors
- Initialization and termination errors

The basis of the Black-Box testing technique or method lies in the selection of appropriate data set. There should be the functional specifications in order to check for normal and abnormal behavior of the software. It is one of the well-researched and practiced classes of testing techniques among the academicians and developers. A brief survey of literature reveals a large number of observations with regard to it. A pertinent selection of those observations include following:

- It should make use of randomly generated inputs but eliminate any guesswork by the tester.
- Input data outside of the specified input range should be tested to check the robustness of the program.
- Boundary cases should be tested to make sure the highest and lowest allowable inputs to produce proper output.
- Whenever numeric data is to be input, the number zero should be tested.
- With real time systems, stress testing should be performed additionally.
- Crash testing should be performed to see what it takes to bring the system down.

- Test monitoring tools should be used to track already performed tests to avoid repetition and to aid in the software maintenance.
- Finite state machine models can be used as a guide to design functional tests.
- Test should be performed in the following general order:
  - (i) clean tests against requirements;
  - (ii) additional structural tests for branch coverage, as needed;
  - (iii) additional tests for data-flow coverage as needed;
  - (iv) domain tests not covered by the above;
  - (v) special techniques as appropriate--syntax, loop, state, etc; and
  - (vi) any dirty tests not covered by the above.

Black Box testing occurs from the viewpoint of an end user and finds bugs such as incorrect functions, interface problems, and database errors. These functional tests examine the observable behavior of software as evidenced by its outputs, without any reference to internal functions. The purpose of Black Box testing is to verify whether a component conforms to specification. It is absolutely mandatory for acceptance testing. The customer must be able to understand these tests to insure whether or not the contract requirements have been met. Black Box technique has several advantages including the following:

- It is more effective on larger units of code than glass box testing.
- It is normally balanced and unprejudiced as the tester and developer are independent of each other.
- It normally determines the quality of the software. Hence may be used to create the quality criteria from this point of view from the beginning.
- Structure of the program is immaterial. Software is subjected to a full range of inputs, and outputs are verified for its correctness.
- It does not require the complex technical skills among the testers.
- There is no need of having detailed functional knowledge of system to the tester.
- It is sequentially performed from end users' point of view.
- It helps in identifying the vagueness and contradiction in functional specifications.
- Test cases can be designed as soon as the functional specifications are complete.
- Professional testers no not need any knowledge of implementation, even the specific programming languages.

It is evident that Black Box testing is concerned only with testing the specification. It cannot guarantee that all parts of the implementation have been tested. Thus Black Box testing is testing against the specifications and will discover faults of omission, indicating that part of the specification has not been fulfilled. In spite of several advantages, Black Box techniques have been blamed to have several disadvantages including the following:

- It is very difficult and challenging to design test cases without having clear functional specifications.

- It is very difficult to identify tricky inputs, since the test cases are not developed based on internal specifications.
- It is very difficult to identify all possible inputs in limited testing time.
- Generating good test cases is a slow and difficult activity.
- There is a chance to have unidentified paths during Black Box testing.
- It is not possible to test every possible input stream.

Black-Box testing is also known as specification-based testing. It refers to test activities using specification-based testing methods and criteria to discover program errors based on program requirements and product specifications. It focuses mainly to find out following type of errors:

- Specification-Based Function Errors
- Specification-Based Component or System Behavior Errors
- Specification-Based Performance Errors
- User-Oriented Usage Errors
- Black-Box Interface Errors

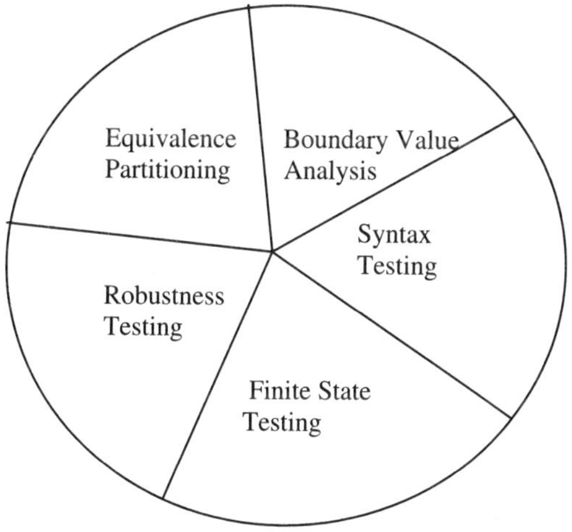


**Figure 4.2: Black-Box Techniques**

Black-Box approach considers only inputs and outputs as a basis for designing test cases. However, due to finite time and resources, an exhaustive test of all possible inputs is not possible. Therefore, it is a goal of a tester by using available resources to produce the test cases that give a maximum number of errors. There are several techniques, as depicted in figure 4.2, that can help to achieve the mentioned goal including the following:

- Equivalence Partitioning
- Boundary Value Analysis

- Robustness Testing
- Syntax Testing
- Finite State Testing

Some of these techniques are more useful at one level than another, while others can be used at any level of test. These techniques are described briefly one by one in the following sections.

## 4.3 EQUIVALENCE PARTITIONING

One of the major issues in software testing is to create test case that is highly likely to uncover program faults. Partitioning divides the input and output domain to generate fewer test cases to represent large input domains and to resolve the issues up to some extent. Partitioning method provides for generating test cases with a better chance of uncovering program faults in particular partitions. To keep down the testing cost, several test cases to test the same aspect of program should not be written. One of the characteristics of a good test case is that it uncovers a different class of errors other than those uncovered by prior test cases.

Equivalence partitioning can be used to reduce the number of test cases that need to be developed by dividing the input domain of a program into classes. As a mathematical property an equivalence class members hold the same property. For each of these equivalence classes, the set of data should be treated the same by the module under test. In equivalence partitioning, test cases should be designed in such a way that the inputs should lie within these equivalence classes. To put it in simpler words, since it is practically infeasible to do exhaustive testing, the next best alternative is to check whether the program extends similar behavior or treatment to a certain set of inputs. If such a set of values can be found in the input domain treat them together as one equivalent class and test one representative test case from this.

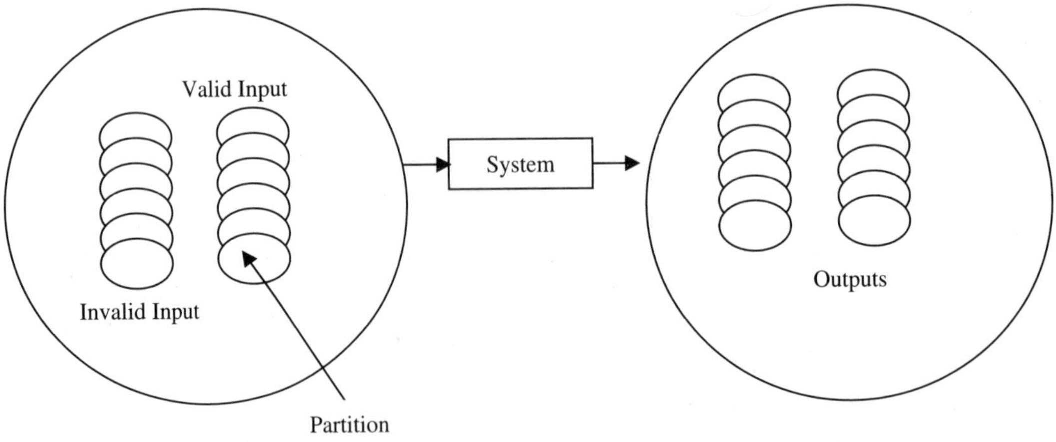


**Figure 4.3: Equivalence Partitioning**

It is a systematic method for identifying sets of interesting classes of input conditions to be tested. Each equivalence class is representative of a large set of other possible tests. Instead of writing test cases for all the elements of the class, one representative test case may be taken to infer the properties of the class. Extra care is needed while picking equivalence classes, and should be ensured that all members of the class behave alike. Figure 4.3 depicts the general view of equivalence partitioning.

### 4.3.1 Scope and Prospects

The equivalence partitioning divides the input domain of the program or module into equivalence classes that have a similar effect on the program. These classes are said to be *Equivalence Classes*. An Equivalence Class is a set of inputs that the program treats identically when the program is tested and is used to represent certain conditions on the input domain. An input condition on the input domain is a predicate over the values of the input domain. Input conditions are typically used to partition the input domain into equivalence classes for the purpose of selecting inputs. During equivalence partitioning, both valid and invalid inputs are considered. A valid input to a program or module is an element of the input domain that is expected to return a non-error value while, an invalid input is an input that is expected to return an erroneous value.

### 4.3.2 Test Case Generation

In order to achieve the objectives of equivalence partitioning, there are two distinct steps to achieve this objective including (i) Identification of equivalence classes; and (ii) Identification of test cases. Identifying the equivalence classes is largely a heuristic process for the given input or external condition. A set of guidelines has been presented below:

(a) If an input condition specifies a range of values then identify

- one valid equivalence class; and
- two invalid equivalence classes.

> **Example:** If a range of values 1....99 is given, then three equivalence classes are required:
>
> - The valid equivalence class {1,.......,99} and;
> - Two invalid equivalence classes {x| x<1} and {x| x>99}.

(b) If an input condition specifies a set of input values and each is handled differently then identify

- one valid equivalence class for each element of the set; and
- one invalid equivalence class.

> **Example:** If the input is selected from a set of say N items then N + 1 equivalence classes are required:
> - One valid equivalence class for each element of the set{$M_1$},……, {$M_N$}; and
> - One invalid equivalence class for elements outside the set {x | x $\notin$ {$M_1$,….., $M_N$}}.

(c) If there is reason to believe that the program handles each valid input differently, then define;
- one valid equivalence class per valid input.

> **Example:** If the input is from a menu then one valid equivalence class for each menu item will be defined.

(d) If the input specifies the number (say N) of valid inputs, then define
- one valid equivalence class for the correct number of inputs; and
- two invalid equivalence classes.

(e) If an input condition specifies a 'must be' situation then identify
- one valid equivalence class; and
- one invalid equivalence class.

> **Example:** The first character of an input must be a numeric character then we require two equivalence classes
> - one valid class: {s| the first character of s is a numeric}; and
> - one invalid class: {s| the first character of s is not a numeric}

(f) If elements in an equivalence class are handled differently by the program, then split the equivalence class into smaller equivalence classes. One straightforward identification may be simple values, common values, extreme values and typical values.

## 4.4 BOUNDARY VALUE ANALYSIS

Programmers often make mistakes on the boundaries of the equivalence classes. So, there is a need to focus the testing process at these boundaries. This type of testing is called Boundary Value Analysis (BVA). It guides to create test cases at the 'edge' of the equivalence classes.

Boundary value is 'a data value that corresponds to a minimum or maximum input, internal, or output value specified for a system or component'.

Boundary value analysis is one of the Black Box techniques. It is based on the observation that faults tend to cluster around 'boundaries' in programs. The boundaries partition the input domain. Boundary value analysis assumes that the program partitions the input into a set of domains in which the program's behavior is similar. Therefore, if an element from some input domain produces an error then a similar error will be produced for all elements of that input domain. It is also assumed that if a test case fails to produce an error then all other elements in the domain will fail to produce an error. On the basis of this, an attempt can be made to cover the structure of the program's input, not the program's structure. This will requires only knowledge of program's specification and not its syntax. This assists to start thinking about test cases, as soon specifications are available.

### 4.4.1 Scope and Prospects

Boundary value analysis is a methodology for designing test cases that concentrates software-testing effort on cases near the limits of valid ranges. It refines equivalence partitioning and generates test cases that highlight errors better than equivalence partitioning. Equivalence partitioning requires only a single test case to be chosen while BVA requires one or more test cases to be chosen.

Basic idea is to concentrate software-testing efforts at the extreme ends of the equivalence classes. Errors are most likely to occur at a point where input values change from valid to invalid. Two steps need to be performed during boundary value analysis including identification of equivalence classes, and designing test cases. Following advantages have been observed for using Boundary Value Analysis:

- It has high probability of finding bugs, against a small number of test cases.
- It tests the boundary value and beyond the limits.
- It focuses attention to exception handling.
- It is one of the most effective methods for designing test cases that reveal common errors made in programming

In spite of the several advantages, there are some disadvantages of using Boundary value analysis. Few of these disadvantages are listed as follows:

- It has a very limited focus.
- It works the best when the program is a function of several independent variables that represent bounded physical quantities.

### 4.4.2 Test Case Generation

Boundary value analysis is not as simple as it sounds. Boundary conditions may be subtle and difficult to identify, especially in finding test cases for boundary conditions in the output domain. Several boundary conditions exist in the literature. Following conditions are be kept in mind while going in for Boundary Value Analysis:

- If an *input* condition specifies a range of values, then construct valid test cases for the ends of the range, and invalid input test cases for input points just beyond the ends of the range.
- If an *input* condition specifies a number of values, construct test cases for the minimum and maximum number of values and one beneath and beyond these values.
- If an *output* condition specifies a range of values, then construct valid test cases for the ends of the output range, and invalid input test cases for situations just beyond the ends of the output range.
- If an *output* condition specifies a number of values, construct test cases for the minimum and maximum number of values and one beneath and beyond these values.
- If the input or output of a program is an ordered set (e.g., a sequential file, linear list, table), focus attention on the first and last elements of the set.
- Use your intelligence to search for other boundary conditions.

If the condition is a range of values, create valid test cases for each end of the range and invalid test cases just beyond each end of the range. For example, if a valid range is -9,99 through 9,99, write test cases that include the following:

- Valid test case: -9,99,
- Valid test case: 9,99,
- Invalid test case: -10,00
- Invalid test case: 10,00

A similar rule applies where the, condition states that the number of values must lie within a certain range. For this, select two valid test cases, one for each boundary of the range, and two invalid test cases, one just below and one just above the acceptable range. Then design tests that highlight the first and last records in an input or output file, and look for any other extreme input or output conditions, and at last, generate a test for each of them.

## 4.5 ROBUSTNESS TESTING

It is evident from literature survey that some programs were testing software to only the minimum extent to ensure that the required functionality was met. Very limited or no testing was performed to ensure that the system could handle unexpected user input or system failures. The value of detecting errors early in the development cycle, both in terms of cost and schedule, is well known. Finding and fixing a software problem after delivery is often several times more expensive than finding and fixing it during the requirements and design phase. This statement is as true for robustness problems as it is for functionality defects. The sooner these types of errors can be found and corrected, the less impact there is on the program cost and schedule. If these types of problems aren't discovered during system testing, they can manifest themselves during operations in ways that can be very difficult to diagnose and fix, disrupting operations and causing very expensive troubleshooting and repair activities.

Robustness can be defined as the ability of software to keep intact the acceptable behavior. This acceptable behavior is expressed in terms of robustness requirements. This kind of feature is required for software critical applications because in these applications, the execution environment cannot be fully foreseen during the development time. Robustness of software is a user-oriented quality requirement specifying the degree to which an application shall continue to function properly under abnormal circumstances. It is a quantitative quality factor measuring the actual degree to which an application continues to properly function under abnormal circumstances. Robustness includes the proper handling of the following:

- Invalid input, either by human actors or other applications.
- Failure of hardware components
- Failure of software components
- Failure of external systems on which the system depends.

Though, robustness testing has been less studied in the field of software than hardware community, several approaches have been proposed in literature to automate software robustness. Robustness is the degree to which a system or a component can function correctly in the presence of invalid inputs or stressful environmental conditions. Testing reveals robustness errors in successful test cases and is not able to uncover any errors in the unsuccessful test case. Robustness testing is the system strength testing of an integrated, Black-Box application against its robustness requirements, that is, immunity to vulnerability to faults.

### 4.5.1 Scope and Prospects

Several approaches have been proposed to automate software robustness testing. First approach is particularly adequate to test very large system where neither a specification nor even the source code is available. This approach consists of generating random inputs obtained by considering only the input domain definition. In another approach, when the source code of the system is available, the generation of relevant inputs to test its robustness can be improved. Moreover, some techniques may also rely on some abstract specification of the system behavior to select the test inputs. It is the case for instance in the so-called FOTG (Fault-Oriented Test Generation) approach. Even if this approach seems well adapted to fault-tolerant protocols it only deals with single faults and uses rather simple specification formalism.

One of the typical goals of robustness testing is to cause the application to fail under invalid conditions so that the underlying defects can be identified, analyzed, fixed, and prevented in future. Literature survey reveals several goals and objectives of using robustness testing. Robustness testing, more specifically, is used to achieve the following.

- To validate the application to determine if it fulfills its robustness requirements.
- To help ensure that the application will not fail catastrophically when the unexpected occurs by determining how well it handle failures of its hardware components, software components and dependent applications.
- To cause failures concerning the exceptional operational requirements and robustness requirements.
- To determine the extent to which the application is ready for launch.
- To provide input to the defect trend analysis effort.

One of the major advantages to the robustness testing is that much of this testing can be performed on both commercial off-the-shelf (COTS) software and customized software. It is often difficult to determine the quality of COTS products, and this type of robustness testing can at least provide a top-level indication of the ability of COTS products to detect and handle situations that can produce errors.

### 4.5.2 Test Case Generation

Robustness testing is a type of software testing that attempts to ensure that a software system performs in an acceptable fashion when it is provided with anomalous input or placed in an inappropriate execution environment. It can be directly related to the process of hardware and software fault injection. Moreover, test cases can be generated in the light of the following:

- The iterative and incremental development cycle implies that robustness testing is regularly performed in an iterative and incremental manner.
- Robustness testing must be automated if adequate regression testing is to occur.
- Robustness testing can elicit failures prior to launch.
- To the extent practical, reuse functional test cases as robustness test cases.
- Robustness testing may overlap slightly with usability testing in that disaster recovery may involve manual procedures performed by members of the operational organization.

## 4.6 SYNTAX TESTING

Syntax testing offers a different approach to testing. It focuses on building test cases based on mutation of the syntax of the input interfaces. Thus it is particularly good at revealing weaknesses in the area of input validation. Through syntax testing it is possible to test that valid input strings are accepted, bad strings are rejected and that no strings cause the implementation to fail. In syntax testing, the test cases, that is, the input to the software, are created based on the specifications of languages understood by the interfaces of software. Automated syntax testing requires a formal description of the input language in machine-readable format. Tester must create a specification for the language if the language is hidden.

The most common form of the specifications is BNF (Backus-Naur form) and regular expressions. Both notations are used to define context-free grammar languages. A sentence is a sequence of bytes, which are arranged according to the rules of the language. Context-free languages have traditionally been used in compilers of various sorts to create parsers for input sentences. In syntax testing, a context-free language is the base used to generate sentences. These sentences are then injected, into the software being tested to see if it accepts them, rejects them or fails to process them altogether.

### 4.6.1 Scope and Prospects

The selection of test cases in syntax testing could start with single-error sentences. This is likely to reveal most of the faults assuming that faults are mutually independent and they are triggered by one error in a sentence. Several types of errors under this category are known today, including ones briefly described below:

**Syntax Errors:** Syntax errors violate the grammar of the underlying language by removing an element, adding an extra element, and providing the elements in wrong order. Syntax error may be at op-level, intermediate level or at field level in the grammar hierarchy.

**Delimiter Errors:** Delimiters mark the separation of fields in a sentence, and the violations of the rules governing the placement and the type of characters that must appear as separators between fields are because of the delimiter errors. Delimiter errors can be created by omitting, duplicating, replacing by other characters or by not respecting the size of fixed length fields.

**Syntax Value Errors:** Illegal values for a given field that cannot be processed by the computing platform, cause value errors. Value errors can be created by exploiting dependencies between the values of the fields, for instance when the value or the type of a field is normally computed or determined from the values of other fields. These errors can also be generated by using values on the border or just bordering the specified ranges of values.

**Field Value Errors:** An illegal field in a sentence is called field-value error. Normally, a field value has a range or many disjoint ranges of allowable values. Field errors can include values, which are one-below, one-above and totally out-of-range. Values exactly at the range boundary should also be checked.

**State Dependency Error:** Not all sentences are acceptable in every possible state of a software component. A state dependency error is generated by inputting a correct sentence during an incorrect state.

### 4.6.2 Test Case Generation

There are three possible kinds of incorrect actions including (i) Recognizer does not recognize a good string; (ii) Recognizer accepts a bad string; and (iii) Recognizer crashes during the attempt to recognize a string.

Figure 4.6.1 shows an example of syntax testing. Following methodologies are generally followed during syntax testing:

- Identify the target language or format.
- Define the syntax of the language, formally, in a notation such as BNF (Backus-Naur Form).
- Test the normal conditions by covering the BNF syntax graph of the input language.
- Test the garbage conditions by testing the system against invalid data.

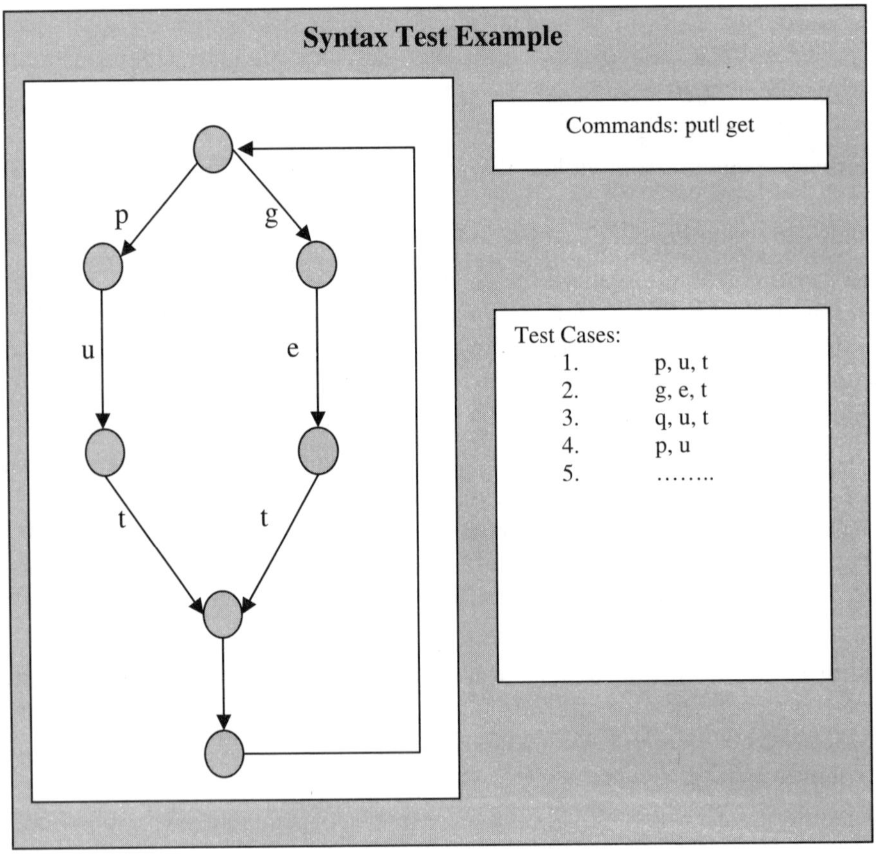


**Figure 4.6.2: Syntax Test example**

Following testing strategies may be adapted to generate test cases during syntax testing:

- Create one error at a time, while keeping all other components of the input string correct.
- Once a complete set of tests have been specified for single errors, do the same for double errors, then triple errors.

- Focus on one level at a time and keep the level above and below as correct as you can.

## 4.7 FINITE STATE TESTING

In the mid-fifties, finite state machine models were used to support automata and switching theory for testing hardware logic. Now a days, it is being used for testing menu-driven applications and to represent abstract data types in object modeling and testing. Finite state machines are represented by directed graphs with the states as nodes and the transitions between states as directed edges. States and Events are used in this approach to specify the behavior of a class in object-oriented programming. A state is an abstraction of the data attributes of a class. A set of attribute values is grouped together into a state according to the properties that affect the gross behavior of the class. Therefore, the state of a class can be defined by predicate conditions on the values of the data attributes of the class. The states of a class include an initial state and a final state.

### 4.7.1 Scope and Prospects

Fine state testing is suitable for object-oriented software because of its features like encapsulation, and may be used at a high-level for Black Box testing. For classes, data members and member functions are encapsulated in one class. Therefore, the value of the data members may be used to determine the states of an instance of the class. Member function may be used to manage data and represent the state transition in state diagram.

In simple class modeling and testing, at most one object instance can exist at any given time. The model generalizes if the states are extended to include the number of object instances in existence at any given time. If the class model includes specification of error conditions, then the states of the class include an error state, s. The error state is the target state for any transition whose event is a member function of the class that raises an error condition. The model generalizes if each error condition is considered to be a distinct error state.

### 4.7.2 Test Case generation

A transition is composed of three items, an event, a guard, and an action. An event can be represented as a call to a member function of the class. A guard is a condition that must be satisfied in order for the transition to occur. It can be a predicate condition on any of the data attributes of the class or on any of the input parameters of the function call. Action will be invoked whenever a guard condition evaluates to TRUE. An action is a set of operations that are performed when the transition occurs. The action can reference or manipulate any of the data attributes of the class or any of the output parameters of the calling function. Whenever an event is received in terms of a message or by invoking a function, the next state of the object depends on the current state as well as the content or action of the event. A change of state caused by an event is called a transition.

A simple class state model with at most one object instance of a class C can exist at any one time and can be represented by the algebraic tuples as follows:

M = (V, F, S, T) where;

V: Finite set of the data attributes of C

F: Finite set of member functions of C

S: Finite set of states of C, i.e. S = {s: s = (predV)} where predV is a predicate on the data attributes in V

T: Finite set of transitions, i.e. T = {t : t = (source, target, event, guard, action)}, where source in S, target in S, event in F, guard is a predicate on attributes in V or parameters of functions in F, and action is a computation on attributes in V or parameters of functions in F.

A simple class state model can be represented as a directed graph with S as the set of nodes and members of T as labeled arcs from the source node of T to the target node of T.

## 4.8 CONCLUSION

Black Box testing of software is a kind of testing where functionality, ignoring the internal workings and structure of the software are being tested. Whenever a Black Box testing is applied to software engineering, the tester should know the legal inputs and the expected outputs. Testers may not know how the program actually arrives at the desired outputs. Such testing implies that the selection of test data as well as the interpretation of test results is performed on the basis of the functional properties of a piece of software. The term Black Box indicates that the internal implementation of the program being executed is not examined by the tester. One of the basic ideas of the Black Box testing is to track which tests have already been performed and the outputs of these tests to avoid repetition. It helps in identifying the vagueness and contradiction in functional specifications.

On the other hand, it is very difficult and challenging to design test cases without having clear functional specifications. Boundary value analysis is a Black Box technique, and is based on the observation that faults tend to cluster around 'boundaries' in programs. Generally it chooses test cases on or around a boundary in order to explore situations around boundary. Boundary value analysis is one of the most effective methods for designing test cases that reveal common errors made in programming. Robustness is the degree to which a system or a component can function correctly in the presence of invalid inputs or stressful environmental conditions. Robustness testing reveals robustness errors in successful test cases and is directly related to the process of hardware and software fault injection. On the other hand, robustness testing may not be able to uncover any errors in the unsuccessful test case. Equivalence partitioning is a strategy that can be used to reduce the number of test cases that need to be developed. Equivalence partitioning divides the input domain of a program into classes. Syntax testing focuses on building test cases based on mutation of the syntax of the input interfaces and thus, is particularly good at revealing weaknesses in the area of input validation.

## KEY TERMS

Black Box Testing, Boundary cases, clean test, Boundary value analysis, Methodology, Boundary conditions, Test case design, Robustness testing, Equivalence partitioning, Equivalence classes, Syntax testing, Syntax error, Delimiter error, Syntax-value errors, Field value errors, State dependency error, Test case generation, Finite state testing.

## REVIEW QUESTIONS

## OBJECTIVE TYPE QUESTIONS

1. Functional testing can be performed at____________.
   (a) System level
   (b) Unit level
   (c) Both system and Unit level
   (d) None of the above
2. Black Box testing is generally performed from____________.
   (a) Programmers' view
   (b) Designers' view
   (c) End user's view
   (d) None of the above
3. In Black Box testing, tester needs to have____________.
   (a) Knowledge of implementation only
   (b) Knowledge of specific program language only
   (c) Both implementation and specific program language
   (d) None of the above
4. Boundary value analysis is the____________.
   (a) Refinement of equivalence partitioning
   (b) Extension of equivalence partitioning
   (c) Both refinement and extension of equivalence partitioning
   (d) None of the above
5. Equivalence partitioning requires choosing only____________.
   (a) Single test case

(b) One or more test cases

(c) Infinite number of test cases

(d) None of the above

6. Robustness testing is the_____________.

(a) System testing of integration

(b) Unit testing

(c) Regression testing

(d) None of the above

7. One of the goals of ____________ is to cause the application to fail under invalid conditions to detect underlying defects.

(a) Regression testing

(b) System testing

(c) Integration testing

(d) None of the above

8. Robustness testing should be performed_____________.

(a) In later phases

(b) In early phases

(c) Regularly in an iterative and incremental fashion

(d) None of the above

9. In an equivalence partitioning, test cases should be designed in such a way that the input should lie_____________.

(a) On the boundary class

(b) Within the equivalence class

(c) Outside the boundary class

(d) None of the above

10. During equivalence partitioning_____________.

(a) Only valid inputs are considered

(b) Only invalid inputs are considered

(c) Both valid and invalid inputs are considered

(d) None of the above

## SHORT ANSWER TYPE QUESTIONS

1. How can overall cost of testing be reduced?
2. List the ways to classify the software testing techniques.

3. How can software be tested without knowledge of internal workings of the software?
4. List the possible type of the legal inputs to programs under Black Box testing.
5. In case of Black Box testing, how selection of test data and interpretation of test result is performed?
6. Define functional testing ans explain with an example.
7. Describe the need and importance of Black Box testing.
8. Elaborate Black Box testing strategy.
9. List out various advantages and disadvantages of Black Box testing.
10. How can you check the robustness of the program during Black Box testing?
11. How can you verify for the highest and lowest allowable inputs to produce proper output in case of Black Box testing?
12. Which type of testing should be performed during stress testing?
13. What do you mean by crash testing? When to perform crash testing?
14. When and how can you use test monitoring tools?
15. Black Box testing does not require the technical skill of testers. Comment.
16. How can Black Box testing help in identifying the vagueness and contradiction in functional specifications?
17. When should test cases be designed for Black Box testing?
18. In Black Box testing, writing test cases are slow and difficult, why?
19. What is boundary value analysis?
20. What do you mean by boundaries in program?
21. Write the methodology used in boundary value analysis.
22. How does boundary value analysis deviate from equivalence partitioning?
23. How can you identify equivalence classes? Explain with an example.
24. How a test case can be designed for equivalence class?
25. Why boundary value analysis generally chooses test cases on or around a boundary?
26. Distinguish between boundary value analysis and equivalence partitioning with reference to the number of test cases.
27. How boundary value analysis focuses attention to exception handling?
28. 'Boundary value analysis is one of the most effective methods for designing test cases'. Justify.
29. What is robustness testing?

30. How can you identify robustness requirements of a software?
31. What type of Black Box testing technique can be used to determine the extent to which the application is ready for launch?
32. List three advantages of robustness testing.
33. Describe the similarities between reuse function test case and robustness test case.
34. Describe the aim and objectives of equivalence partitioning testing.
35. Discuss on the extra care needed while choosing equivalence classes.

## DESCRIPTIVE QUESTIONS

1. Testers and programmers can be independent of one another in Black Box testing.Justify.
2. Why should not Black Box testing be performed by the author of the program who knows too much about the program internals?
3. Finite state machine model can be used to design functional test. Comment.
4. List out at least five strategies associated with Black Box testing not listed in this chapter.
5. How can stress testing help in testing real time system?
6. Testers performing Black Box testing may be non-technical. How a non-technical personal can perform a proper test? Justify.
7. In Black Box testing, since there is no clear functional specification, how test cases are designed?
8. How can boundary value analysis generate test cases that highlight errors better than equivalence partitioning?
9. Review a boundary condition, take a small piece of code and implement it.
10. How the relevant input generation of test in robustness can be improved?

## REFERENCES

1. Van Tassel, Dennie Program Style, Design, Efficiency, Debugging, and Testing. Englewood Cliffs. Prentice Hall. 1974.
2. Beizer, Boris Software Testing Techniques, 2nd Edition. New York. Van Nostrand Reinhold. 1990.
3. Dunn, Robert H. Software Defect Removal. New York. McGraw-Hill. 1984.

4. Quirk, W. J. Verification and Validation of Real-Time Software. New York. Springer-Verleg. 1985.

5. Beizer, Boris Software System Testing and Quality Assurance. New York. Van Nostrand Reinhold. 1984.

6. Bertolino, A. (May 2001). Chapter 5: Software Testing. IEEE SWEBOK Trial Version

7. Boehm, B. W. (1981). Software Engineering Economics. Englewood Cliffs, NJ, Prentice-Hall, Inc.

8. Copeland, L. (2004). A Practitioner's Guide to Software Test Design. Boston, Artech House Publishers.

9. Craig, R. D. and S. P. Jaskiel (2002). Systematic Software Testing. Norwood, MA, Artech House Publishers.

10. Dijkstra, E. W., "Notes on Structured Programming," Technological University Eindhoven T.H. Report 70-WSK-03, Second edition.

11. Galin, D. (2004). Software Quality Assurance. Harlow, England, Pearson, Addison Wesley.

12. IEEE (1986). "ANSI/IEEE Standard 1008-1987, IEEE Standard for Software Unit Testing."

13. IEEE (1987). "ANSI/IEEE Standard 1008-1987, IEEE Standard for Software Unit Testing."

14. IEEE (1990). IEEE Standard 610.12-1990, IEEE Standard Glossary of Software Engineering Terminology.

15. IEEE, "IEEE Standards Collection: Glossary of Software Engineering Terminology," IEEE Standard 610.12-1990.

16. Kaner, C., J. Bach, et al. (2002). Lessons Learned in Software Testing, John Wiley & Sons.

17. Knuth, D. E. (1992). The errors of TeX. Software--Practice and Experience. Literate Programming; CSLI Lecture Notes, no. 27, CSLI. **19:** 607--681.

18. Martin, R. C. (2003). Agile Software Development: Principles, Patterns, and Practices. Upper Saddle River, Prentice Hall.

19. Myers, G. J. (979). The Art of Software Testing. New York, John Wiley.

20. Pressman, R. (2001). Software Engineering: A Practitioner's Approach. Boston, McGraw Hill.

21. Gen. John J. Sheehan. A commander-in-chief's view of rear-area, home-front vulnerabilities and

22. support options. In Proceedings of the Fifth InfoWarCon, September 1996. Presentation, September 5.

23. B.P. Miller, L. Fredrikson, and B. So. An empirical study of the reliability of UNIX utilities. Communications of the ACM, 33(12):32-44, December 1990.

24. B.P. Miller, D. Koski, C.P. Lee, V. Maganty, R. Murthy, A. Natarajan, and J. Steidl. Fuzz revisited: A re-examination of the reliability of UNIX utilities and services. Technical report, University of Wisconsin, Computer Sciences Dept, November 1995.

25. P. Koopman, J. Sung, C. Dingman, D. Siewiorek, and T. Marz. Comparing operating systems using robustness benchmarks. In Proceedings of the 16th IEEE Symposium on Reliable Distributed Systems, pages 72-79, October 1997.

26. Ghosh, M. Schmid, V. Shah. An Approach for Analyzing the Robustness of Windows NT Software. In Proceedings of the 21st National Information Systems Security Conference, October 5-8, 1998, p. 374-382. Crystal City, VA.

27. Ghosh, M. Schmid, V. Shah. Testing the Robustness of Windows NT Software. To appear in the International Symposium on Software Reliability Engineering (ISSRE'98), November 4-7, 1998, Paderborn, GE.

28. J. Richter. Advanced Windows, Third Edition. Microsoft Press, Redmond Washington, 1997. Page 529.

## USEFUL LINKS

http://www.cis.ysu.edu/~kramer/DataStructures/Intro/
http://en.wikipedia.org/wiki/Black_box_testing

http://www.cse.fau.edu/~maria/COURSES/CEN4010-SE/C13/black2.html

## Chapter 5

# White Box Testing

*Never stop testing, and your advertising will never stop improving.*
**David Ogilvy**

## OBJECTIVES

A successful completion of this chapter will enable the reader for the following:

- To understand the term 'White-Box'.
- To appreciate the need, importance and significance of White-Box testing technique.
- To comprehend the role of White-Box testing technique in software engineering.
- To list the advantages and disadvantages of White-Box testing technique.
- To elaborate the scope and objectives of White-Box modeling.
- To define and understand the prospects basis path testing.
- To comprehend the role of flow graph in test case design.
- To elaborate test case design for basis path testing.
- To define the way to calculate cyclomatic complexity.
- To identify the benefits of using graph matrix.
- To elaborate the use of graph matrix in calculating cyclomatic complexity of a flow graph.
- To relate control structure testing including condition coverage, branch coverage, and statement coverage.
- To identify the scope and objectives of code coverage.
- To comprehend the role of data flow testing and loop testing in White-Box testing.
- To realize the need of gray box testing.
- To understand gray box testing process and the methodology used for it.
- To follow the mutation testing process to test software.
- To implement White-Box testing, by generating test cases using different techniques.

## 5.1 INTRODUCTION

For a long time, the majority of the programmers believed that programs are only written to be read by a computer. It was also their opinion that the only way to test a program is to execute it on the machine for which it was designed. This attitude often led towards badly structured, illegible, and nearly unchangeable programs. But what if a programmer knows that he will have to present his product to other professionals? Additionally, such a presentation might disclose some bugs which are otherwise hard to find, resulting in a strong psychological influence on the programmer. He will automatically write programs which are simpler to read, better structured, and easier to maintain. During such a presentation the internal structure of a program is being disclosed with the main goal to detect faults. Therefore such a procedure is in general called a 'White-Box Test'. White-box testing is an important method for the early detection of errors during software development. In this process test case generation plays a crucial role, defining appropriate and error sensitive test data. The evolutionary white-box testing is a promising approach for the complete automation of structure oriented test case generation.

One of the major objectives of White-Box testing is to provide sets of test values to ensure that every statement in the procedure is executed at least once. This includes all statements even those executed only when optional arguments, user-supplied arguments, subroutines, user-action routines, or specific error codes are present. White-Box testing is responsible for providing test cases to ensure that each branch of a decision is executed at least once. In the case of a standard Boolean decision, this typically requires providing two values; however, this number may be much greater in the case of compound or nested decisions. Condition testing requires writing test cases that ensure each condition in a decision takes all possible outcomes at least once, and each point of entry to the program or subroutine is invoked at least once. White-Box testing supplies multiple test values in cases of compound and nested loops. In testing the entry points, remember to invoke any optional routines (either internal or external), as well as error handlers.

White-Box testing of software is based on close examination of procedural details of software components, particularly programs. It allows looking inside a component being tested and creating test cases based on the component's implementation (code). White-box testing is a strong verification technique through which software engineers examine if their code works as expected. In White-Box testing logical paths through the software are tested by providing test cases that exercise specific sequence of instructions, conditions and/or loops. The status of the program may be examined at various points to determine if the expected or asserted status corresponds to the actual status.

Figure 5.1 (a) gives simple White-Box specifications. White-Box testing uses the structure of the code to select test cases. Structure of the code may be used to develop methods that explore the possible execution paths and data dependencies in program in various ways. *Data Flow* criteria may be used to select White-Box test cases by using dependencies between variables and their usage. On the other hand, a *Control Flow* criterion, which uses the possible execution paths in a program, may be used to select White-Box test cases. In order to conduct White-Box testing, followings may be generally required as an input:

- A White-Box testing model and test criteria;
- A White-Box test design and generation method; and
- Program's source code.

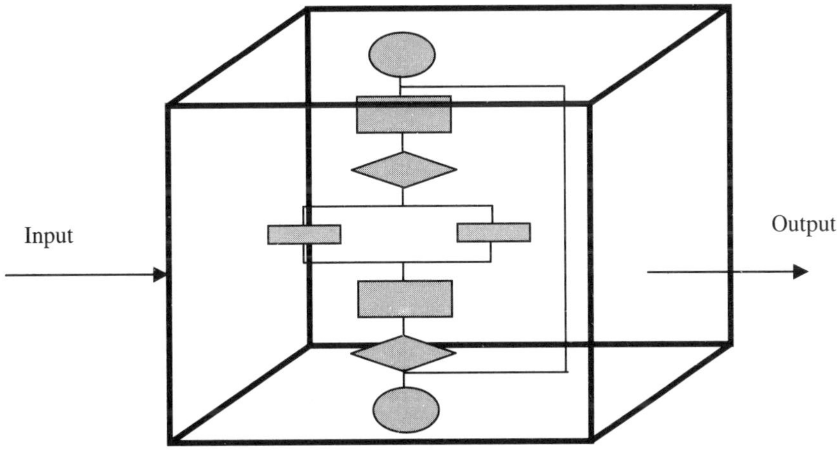


**Figure 5.1(a): A Simple White-Box Specification**

There are many basic facts, which form the basis for White-Box type of testing techniques. A selection from those include following:

- Logical errors and incorrect assumptions are inversely proportional to a path's execution probability.
- It is often believed that a path is not likely to be executed; in fact, reality is often counter intuitive.
- Typographical errors are random; it's likely that untested paths will contain some.

**Table 5.1: White-Box Technique and Knowledge Source**

| **Technique** | **Knowledge Source** |
|---|---|
| White-Box Technique | • High-level design<br>• Detailed design<br>• Control flow graphs |

White-Box testing is also known as structural testing or code-based testing. Table 5.1 shows the knowledge source for White-Box technique. Test cases to be used to test a program using White-Box testing are derived from the program structure, and not from functional specifications. Structural test cases must wait to be derived until the code is written. Structural testing is used to back up functional testing. Figure 5.1(b) depicts White-Box technique, which includes the execution of following:

- All independent paths within a module at least once;
- All logical decisions on their true and false sides;
- All loops at their boundaries band within their operational bounds; and
- All internal data structures to assure their validity.

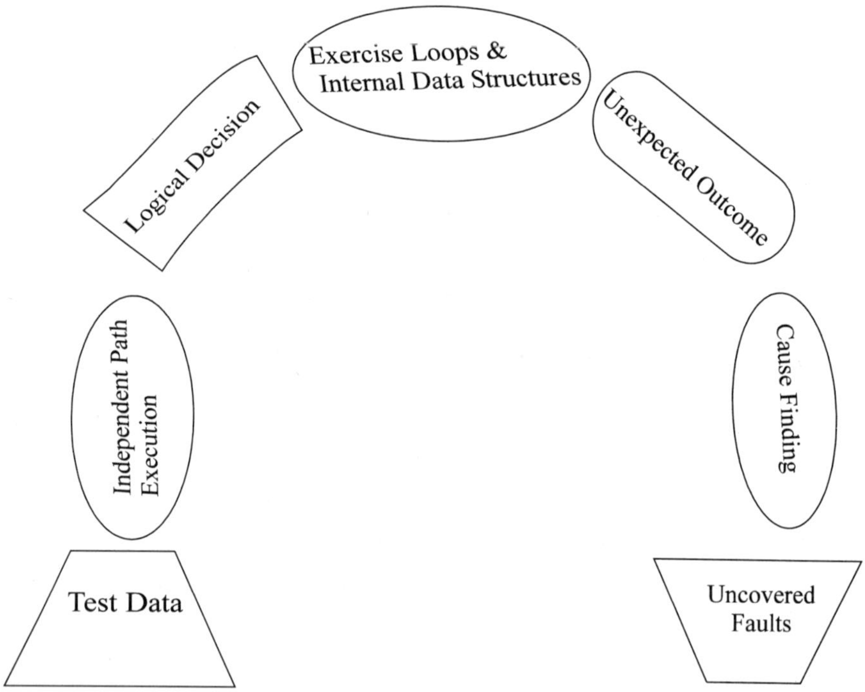


**Figure 5.1(b): White-Box Testing**

## 5.2 WHITE-BOX TECHNIQUES

White-Box testing is defined as performing tests with knowledge of the implementation, such as requiring that every line of code be executed at least once, or that each function be individually tested. Studies have shown that when White-Box testing is added to Black-Box, quality is improved by 20 percent, a significant amount. The major objective of White-Box testing is to focus on internal program structure, and discover all internal program errors. It focuses on program structures including program statements and branches, various kinds of program paths, program's internal logic and data structures, and program's internal behaviors and states. Several advantages of White-Box testing have been observed including the following:

- As the knowledge of internal coding structure is prerequisite, it becomes very easy to find out which type of input/data can help in testing the application effectively.
- It helps in optimizing the code.
- It helps in removing the extra lines of code, which can bring in hidden defects.
- It approximates the partitioning done by execution equivalence.
- It reveals errors in hidden code.
- In White-Box testing, crosschecking design decisions and assumptions against source code may outline a robust design.
- All the possible code pathways can be tested including error handling, resource dependencies, and additional internal code logic/flow.
- It analyzes source code and developing tests based on the implementation details enables testers to find programming errors quickly.
- It ascertains quality of coding work and its adherence to coding standards.

White-Box testing saves time, but it has its disadvantages, too. If several things are being tested at once, and correct answers are being observed, it is a pretty good indication that everything works. If the correct answer is not being observed, then it can't be assured what went wrong. White-Box testing technique requires code modification has been done using interactive debuggers, or by actually changing the source code. This may be adequate for small programs; however, it does not scale well to larger applications. Following disadvantages of White-Box testing have been observed:

- As knowledge of code and internal structure is a prerequisite, a skilled tester is needed to carry out this type of testing, which increases the cost.
- It is nearly impossible to look into every bit of code to find out hidden errors, which may create problems, resulting in failure of the application.
- It is one of the very expensive techniques of testing.
- It requires intimate knowledge of target system, testing tools and coding languages, and modeling.
- It requires specialized tools such as source code analyzers, debuggers, and fault injectors.
- In White-Box testing, the demarcation between developer and testers starts to blur which may become a cultural stress.

A major challenge in testing is to determine a good starting set of test cases that eliminates redundant testing, provides adequate test coverage, allows more effective testing, and makes the most of limited testing resources. White-Box testing techniques may be classified as Traditional White-Box techniques, Object-oriented White-Box technique and Component-Oriented White-Box techniques.

Traditional White-Box techniques are discussed in this chapter, which is further divided into two types as shown in figure 5.1(c), as follows:

- Basis Path Testing
- Control Structure

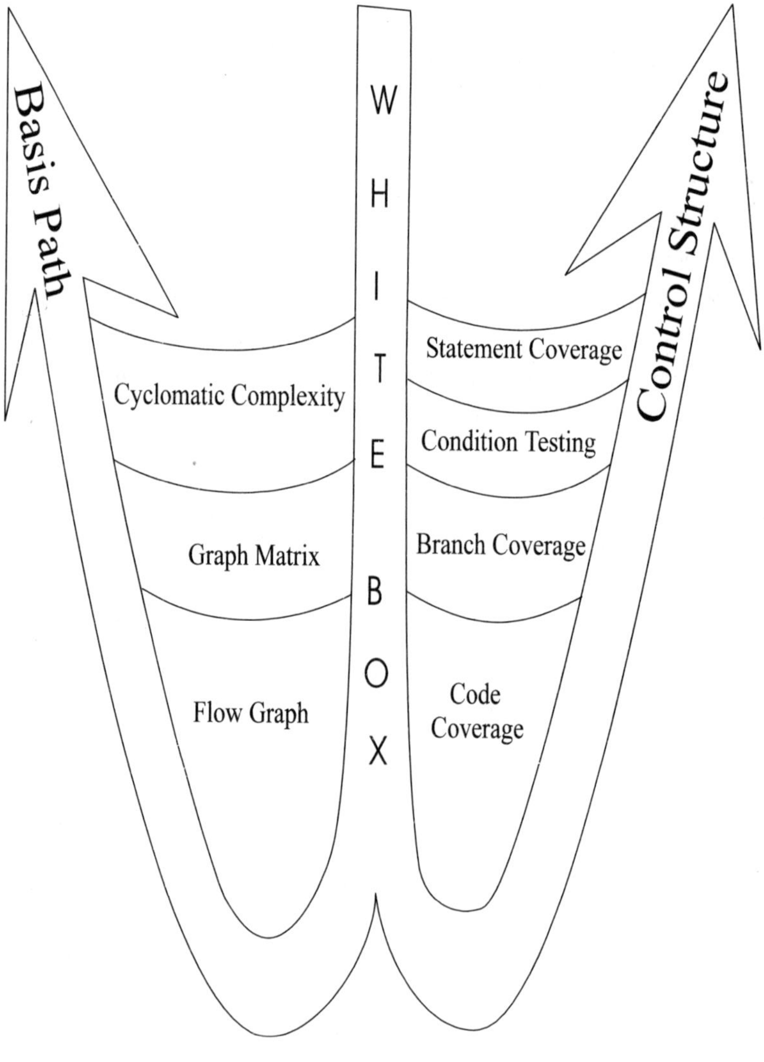


**Figure 5.1(c): Traditional White-Box Techniques**

## 5.3 WHITE-BOX MODELING

The intention in White-Box testing is to ensure that all feasible flow of control paths through a subprogram is traversed for the software is under test. This is not the same as saying that all statements in the subprogram will be executed as it is possible for all statements to be executed but for not all of the possible paths to be traversed. However, the converse is true; if all possible paths through a subprogram are traversed then all statements in the subprogram will necessarily be executed. Software product quality is synonymous with 'fitness-for-purpose and fitness-off-form' and such fitness can be determined through dimensions such as functionality, reliability, usability and maintainability. Every time software is used, it is expected to perform a task in a specific manner. If it does not behave the way it is intended to, it is considered useless.

Software Testing is the process of examining something with the intention of finding errors. Testing can uncover these 'misbehaviors', and therefore, software is tested before it is delivered to a customer. Therefore, there is a need to design an effective set of test cases that enable uncovering maximum errors within tight project schedules. Test case design involves designing a set of test inputs, execution conditions, and expected results for testing an aspect of the software. Its aim is to create an optimal set of test cases to test all aspects of the software. A good test case is traceable to customer requirements, modular, and reusable.

While testing may reveal a symptom of an error, it may not uncover the exact cause of the error. White-Box testing is the testing of the underlying implementation of a piece of software without regard to the specification or external description for that piece of software. The goal of White-Box testing of source code is to identify such items as (unintentional) infinite loops, paths through the code which should be allowed, but which cannot be executed and dead (unreachable) code. White-Box testing is mandatory in software development lifecycle. Several techniques have been identified to generate test cases while using White-Box testing including following:

- Control Flow Graph;
- Cyclomatic Complexity; and
- Graph Matrix.

**Control Flow Graph:** White-Box testing looks inside the objects being tested and uses that knowledge in the testing process. It relies on the structure of a program. Therefore, it is required to define the representation of the program. Flow graph is one of the ways to depict the structure and logical flow of the program.

Any procedural design can be translated into a control flow graph. In a flow graph, lines or arrows are used to represent flow control. These lines are called edges. Circles are used to represent one or more actions and called nodes. A node containing a condition is called a predicate node. Area bounded by the edges and nodes are called regions. Figure 5.3(a)-(f) control flow graph notations for different program structures. Figure 5.3(g) shows flow graph notation for case statement.

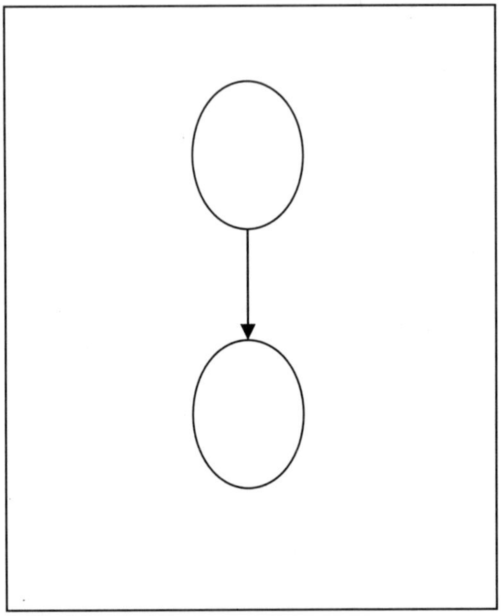

Figure 5.3(a): Flow Graph for Sequence

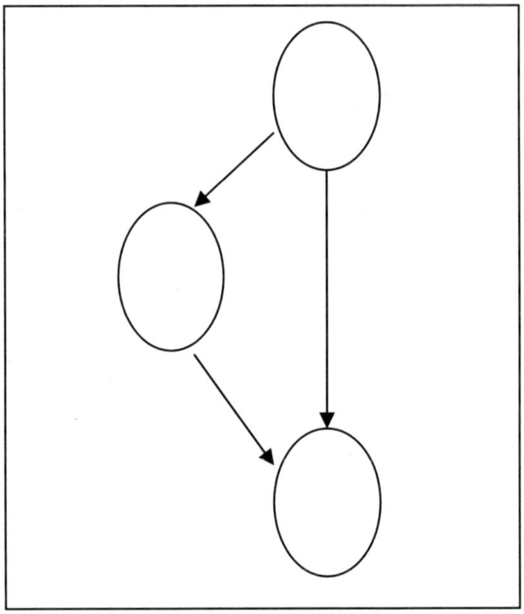

Figure 5.3(b): Flow Graph for Selection

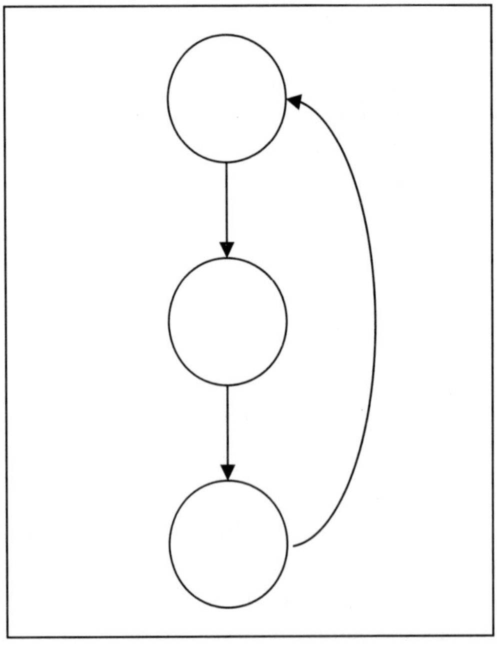

Figure 5.3 (c): Flow Graph for Loop

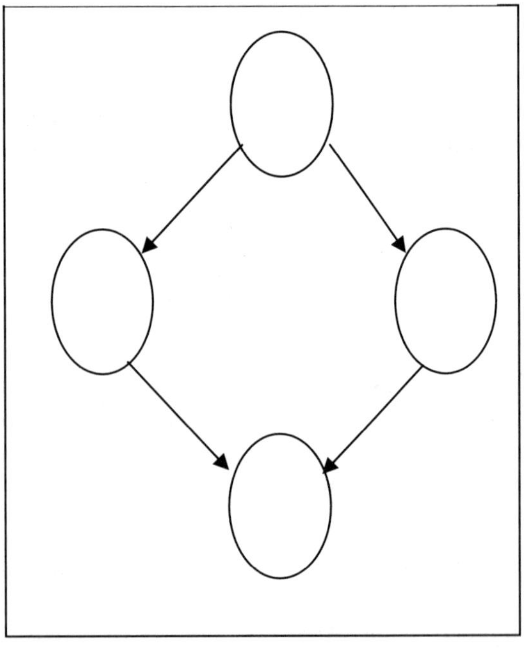

Figure 5.3 (d): Flow Graph for If-Then-Else

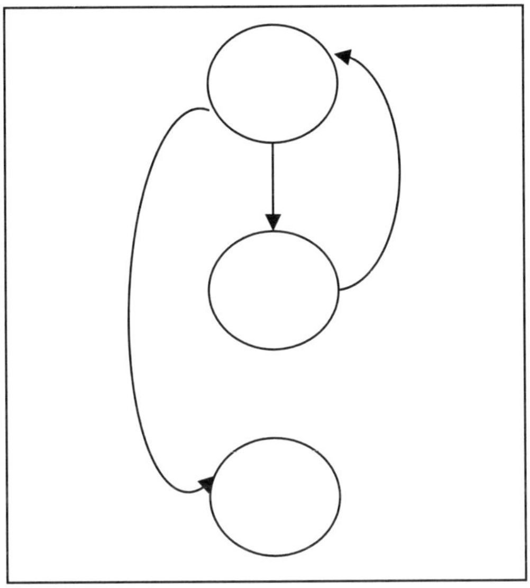

**Figure 5.3 (e): Flow Graph for While Loop**

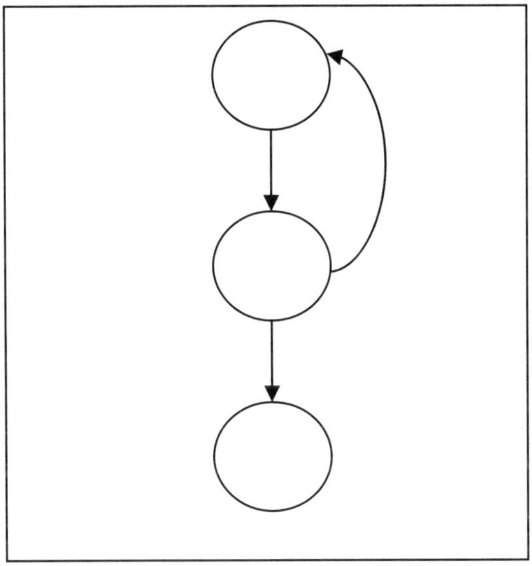

**Figure 5.3(f): Flow Graph for Until Loop**

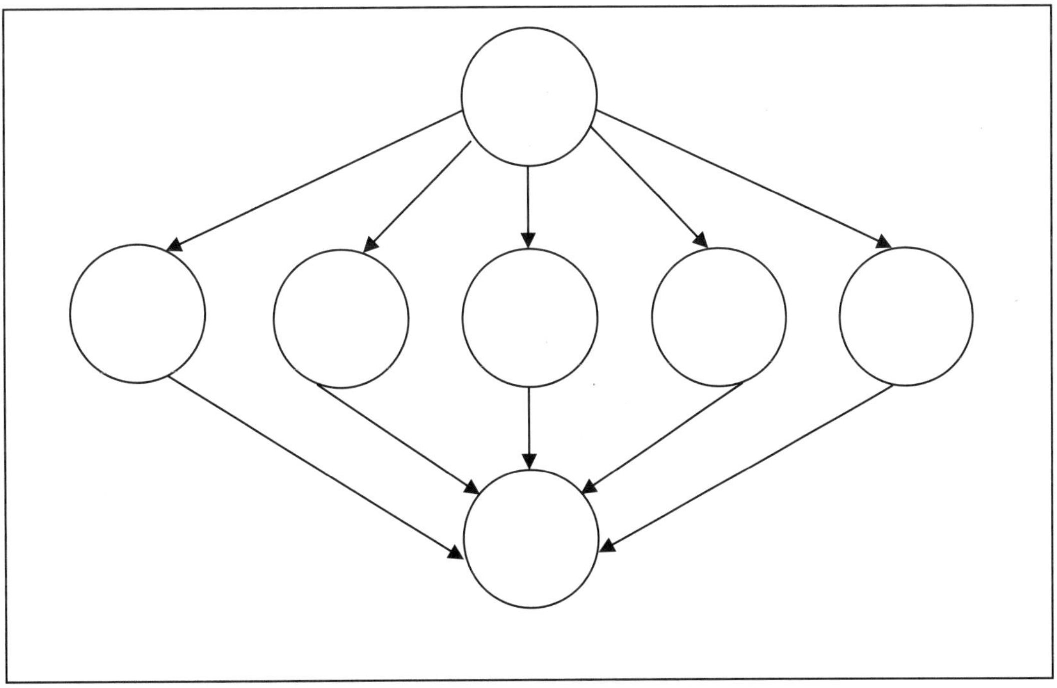

**Figure 5.3(g): Flow Graph for Case Construct**

Figure 5.3(h) shows the flow graph notation for compound Boolean expressions, where at least two predicate nodes and additional arcs are generated.

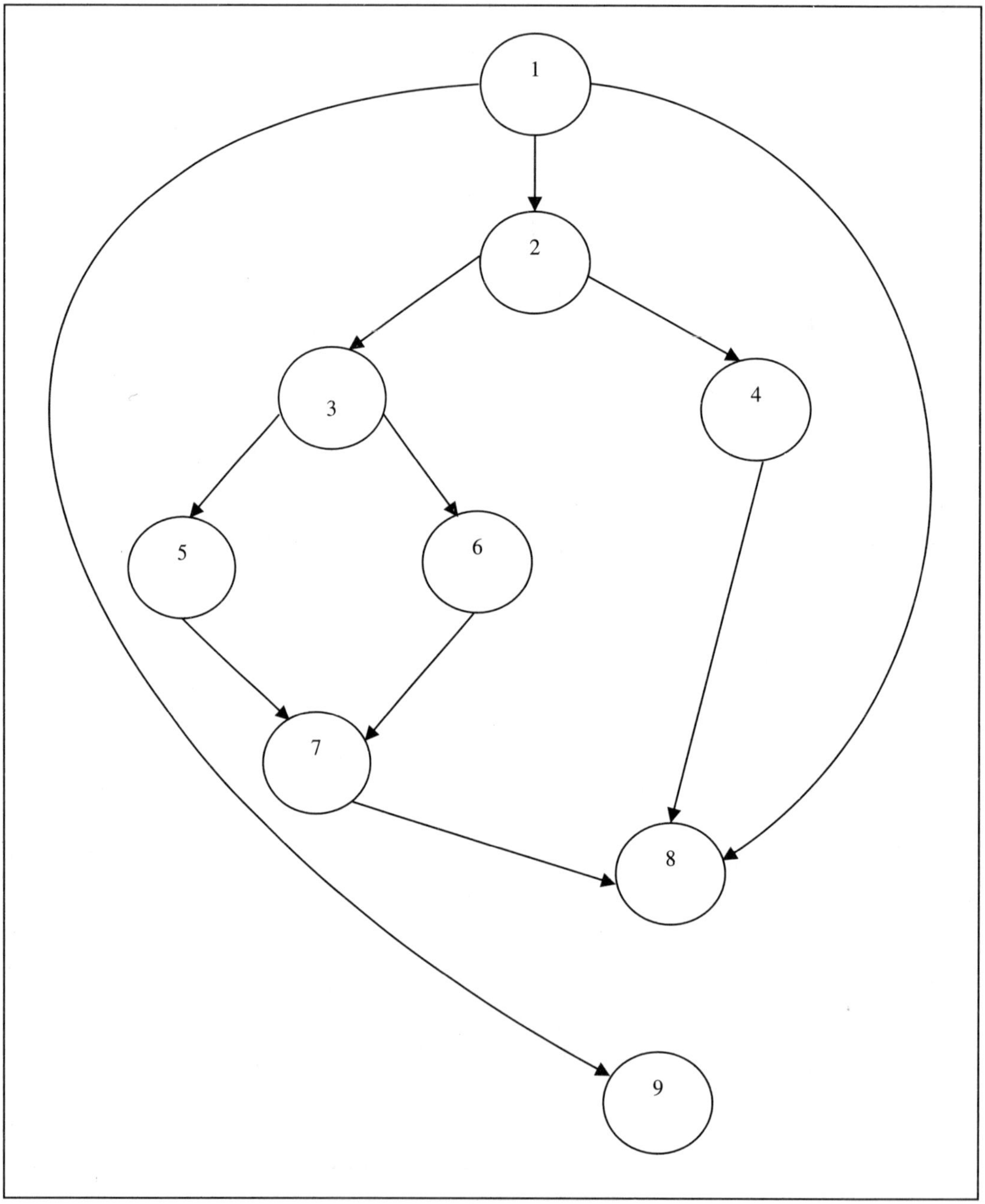


**Figure 5.3(h): Flow Graph for Compound Boolean Expression**

**Cyclomatic Complexity:** The cyclomatic complexity derives a quantitative measure of logical complexity. This value gives the number of independent paths in the basis set, and an upper bound for the number of tests to ensure that each statement is executed at least once. The word 'cyclomatic' comes from the number of fundamental cycles in connected, undirected graphs. It gives the number of independent paths through strongly connected directed graphs. A strongly connected graph is one in which each node can be reached from any other node by following directed edges in the graph. Program control flow graphs are not strongly connected, but they become strongly connected when a 'virtual edge' is added connecting the exit node to the entry node. It is defined for each module to be e - n + 2p, where e and n are the number of edges and nodes in the control flow graph, respectively and p is the number of unconnected parts of the graph. Cyclomatic complexity is also known as v (G), where v refers to the cyclomatic number in graph theory and G indicates that the complexity is a function of the graph. Broadly Cyclomatic Complexity can be used for following:

- To find the number of paths to look for in basis path testing.
- To help pinpoint areas of potential instability.
- To indicate a unit/component's testability, understandability and maintainability.
- To provide a quantitative indication of unit/component's control flow complexity.
- To indicate the effort required to test a unit/component.

Figure 5.3(i) and 5.3(j) show cyclomatic complexities computation for a sequential code and compound Boolean code segments respectively.

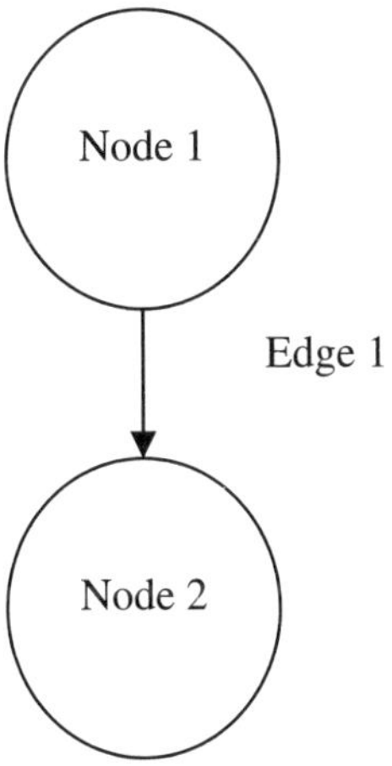


Total Number of Nodes, n= 2

Total Number of Edges, e= 1

Total Number of Unconnected Part of the Graph, p=1

Cyclomatic Complexity of Sequential code= 1-2+2*1=1

**Figure 5.3 (i): Cyclomatic Complexity of Sequential Code Segment**

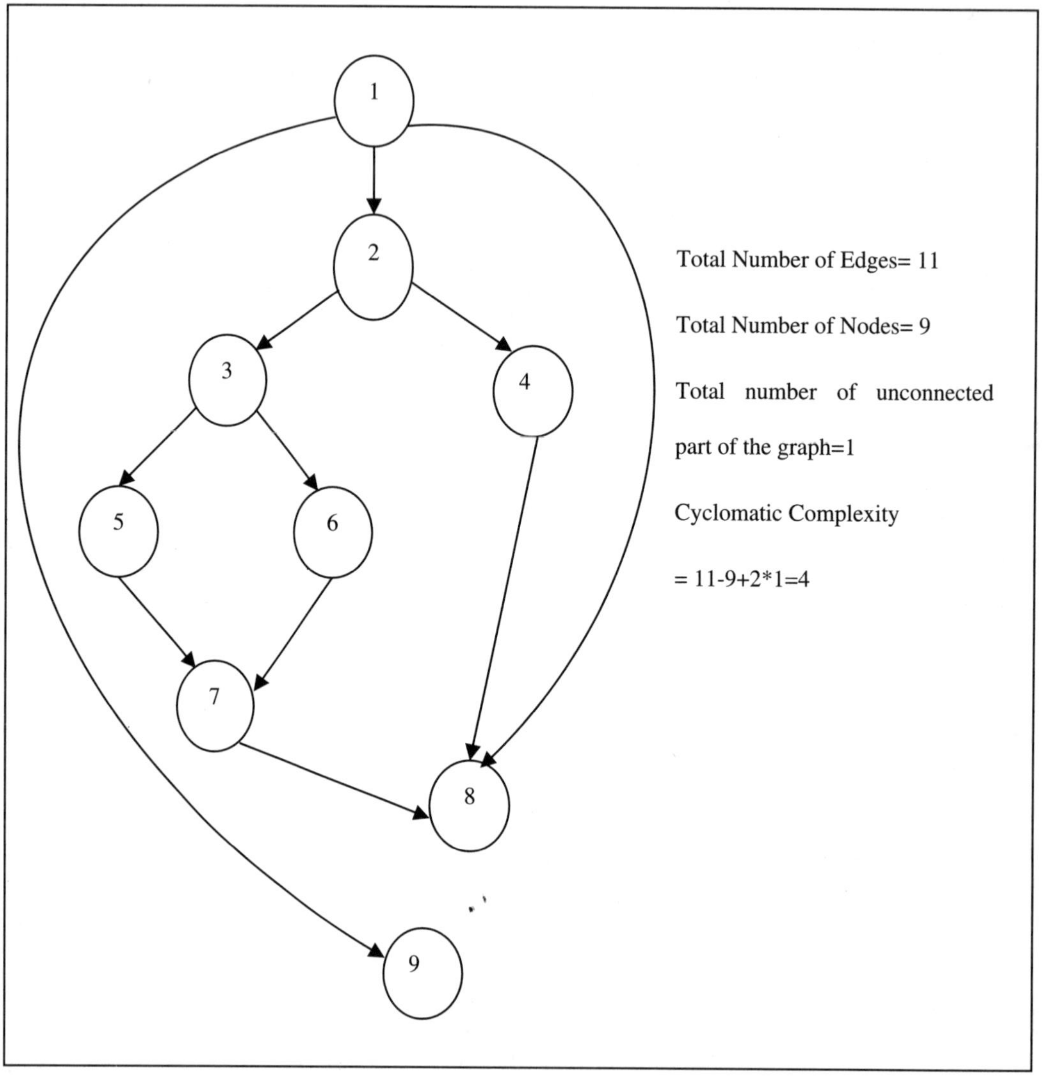


**Figure 5.3(j): Cyclomatic Complexity for Compound Boolean Code Segment**

Another way of calculating Cyclomatic Complexity is to use predicate formula as follows:

**V(G) = Number of Predicate Nodes + 1**

Figure 5.3(k) shows Cyclomatic Complexity calculation using both the formula. Cyclomatic Complexity obtained using both the formula are same. Figure 5.3(l) shows another flow graph and Cyclomatic Complexity for it. But for this flow graph, Cyclomatic Complexity obtained using both the formula is different. This difference is because of the limitation on the predicate nodes formula. Predicate node formula assumes that there are only two outgoing flows for each of such nodes. To adjust for this, predicate node can be split into two sub-nodes as shown in figure 5.3(m).

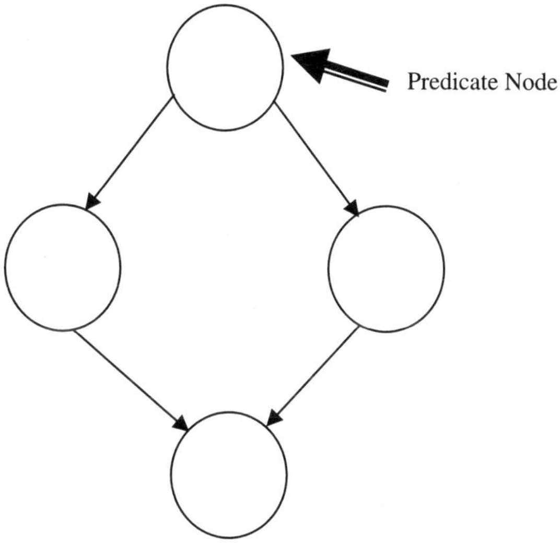


**Using Predicate Formula**

Number of Predicate Nodes=1
V (G)= Number of Predicate Nodes+1
= 1+1=2

**Using V(G)=e-n+2p**

Number of Edges= 4
Number of Nodes= 4
Number of unconnected part of the graph=1
V (G) =4-4+2*1=2

**Figure 5.3 (k): Cyclomatic Complexity Using Both the Formula**

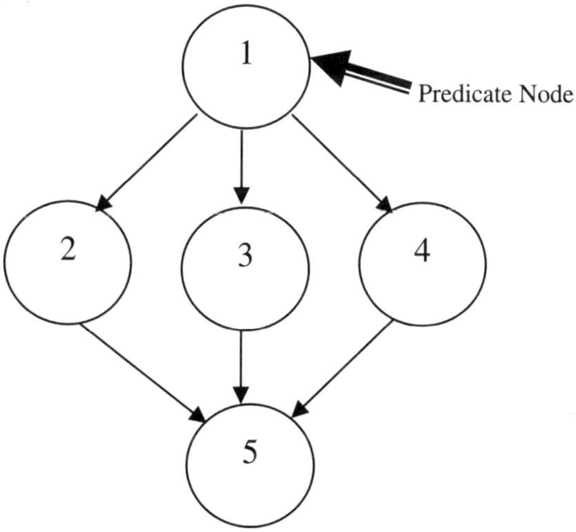


**Using Predicate Formula**

Number of Predicate Nodes=1
V (G)= Number of Predicate Nodes+1
= 1+1=2

**Using V(G)=e-n+2p**

Number of Edges= 6
Number of Nodes= 5
Number of unconnected part of the graph=1
V (G)=6-5+2*1=3

**Figure 5.3 (l): Cyclomatic Complexity Using Both the Formula**

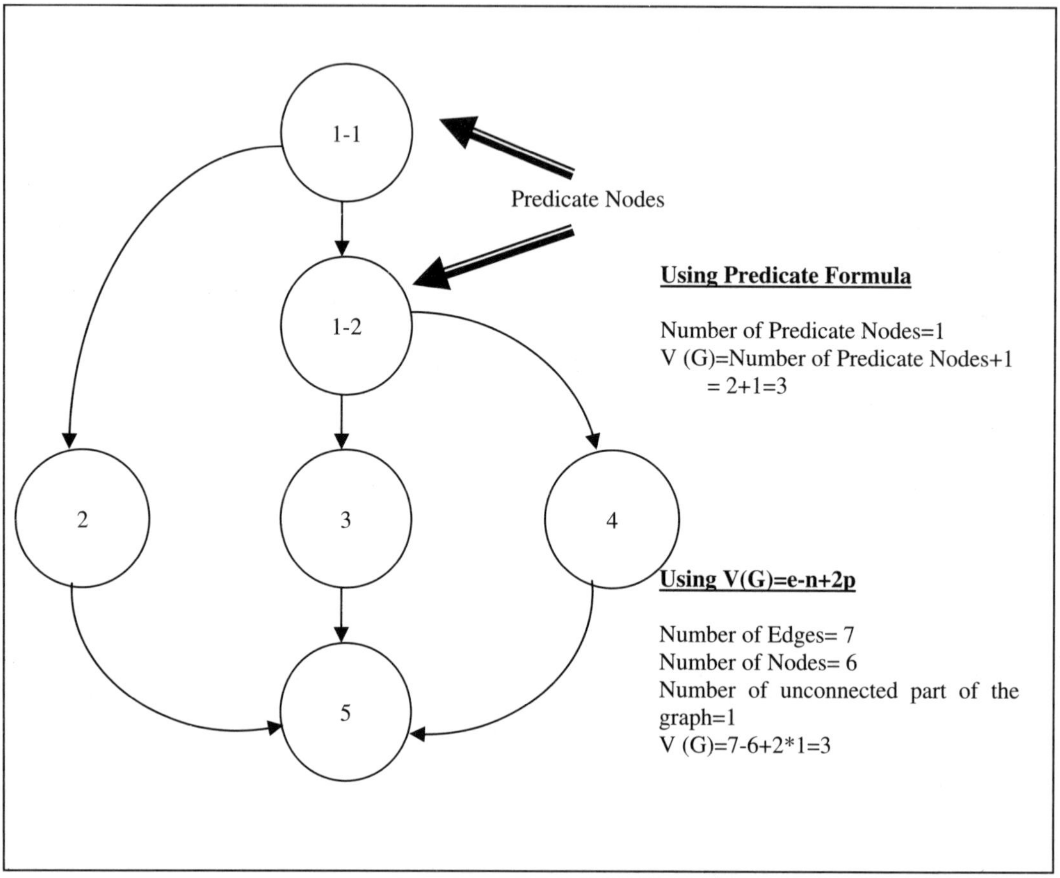


**Figure 5.3 (m): Cyclomatic Complexity Using Both the Formula**

**Graph Matrix:** The procedure for deriving the flow graph and even determining a set of basis paths is amenable to mechanization. To develop a software tool that assists in basis path testing, a data structure called a graph matrix can be quite useful. A graph matrix is a square matrix whose size (that is, number of rows and columns) is equal to the number of nodes on the flow graph. Each row and column corresponds to an identified node, and matrix entries correspond to connections (an edge) between nodes.

Referring to the figure 5.3(n), each node on the flow graph is identified by numbers, while each edge is identified by letters. A letter entry is made in the matrix to correspond to a connection between two nodes. For example, node 3 is connected to node 4 by edge e. Up to this point, the graph matrix is nothing more than a tabular representation of a flow graph. However by adding a link weight to each matrix entry, the graph matrix can become a powerful tool for evaluating program control structure during testing. The link weight provides additional information about control flow. In its simplest form, the link weight is 1 (a connection exists) or 0(a connection does not exist). But link weights can be assigned other, more interesting properties including following:

- The probability that a link (edge) will be executed.
- The processing time taken during traversal of a link.
- The memory required during traversal of a link.
- The resources required during traversal of a link.

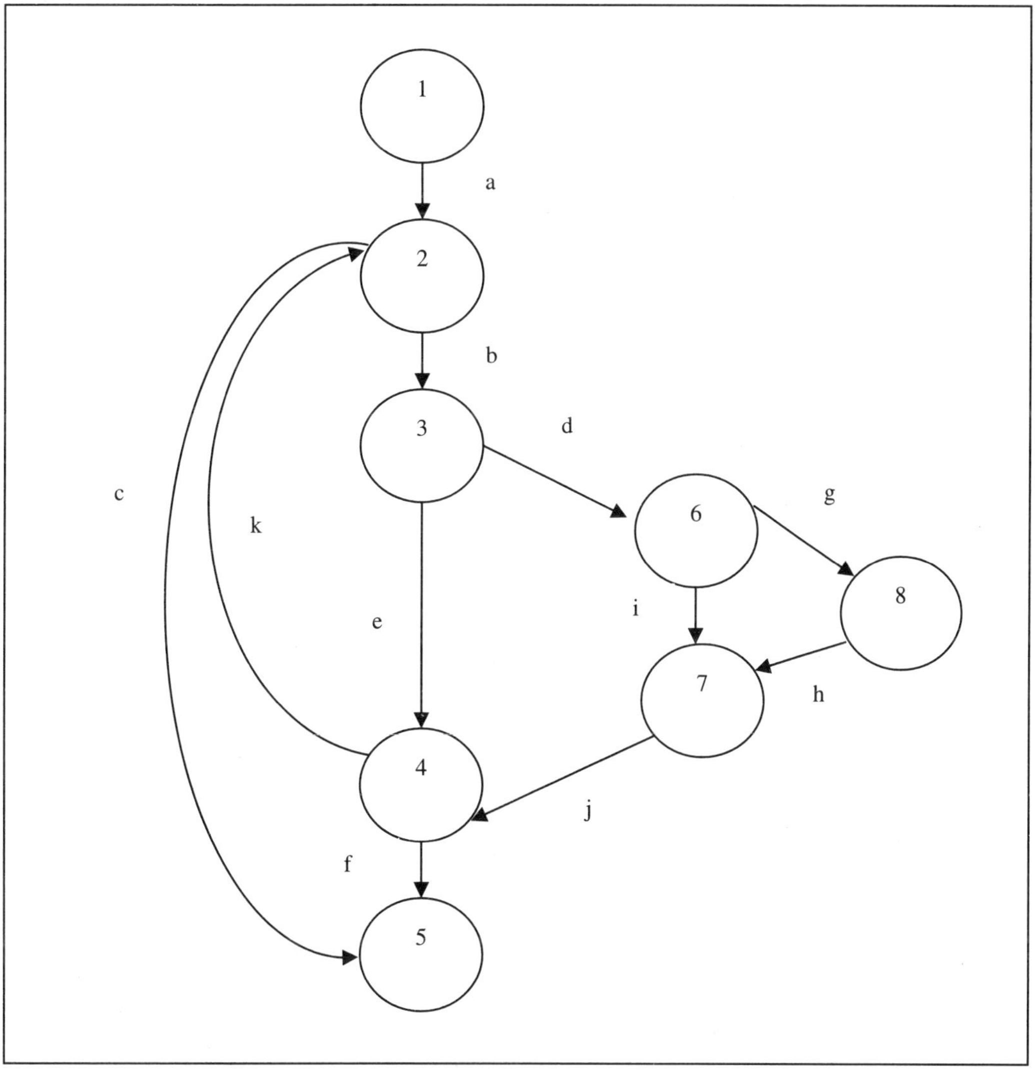


**Figure 5.3(n): Flow Graph**

To illustrate, use the simplest weighting to indicate connections (0 or 1). The graph matrix shown in table 5.3(b) is redrawn in table 5.3(c).

**Table 5.3(b): Graph Matrix for Flow Graph**

| Nodes↓ | Edges→ | | | | | | |
|---|---|---|---|---|---|---|---|
| 1 | | a | | | | | |
| 2 | | | b | | c | | |
| 3 | | | | e | | d | |
| 4 | | k | | | f | | |
| 5 | | c | | | | | |
| 6 | | | | | | i | g |
| 7 | | | | j | | | |
| 8 | | | | | | h | |

**Table 5.3(c): Connection Matrix for the Graph Matrix**

| | Connected to a Node | | | | | | | |
|---|---|---|---|---|---|---|---|---|
| **Node** | 1 | 2 | 3 | 4 | 5 | 6 | 7 | 8 |
| 1 | 0 | 1 | 0 | 0 | 0 | 0 | 0 | 0 |
| 2 | 0 | 0 | 1 | 0 | 0 | 0 | 0 | 0 |
| 3 | 0 | 0 | 0 | 1 | 0 | 1 | 0 | 0 |
| 4 | 0 | 1 | 0 | 0 | 1 | 0 | 0 | 0 |
| 5 | 0 | 0 | 0 | 0 | 0 | 0 | 0 | 0 |
| 6 | 0 | 0 | 0 | 0 | 0 | 0 | 1 | 1 |
| 7 | 0 | 0 | 0 | 1 | 0 | 0 | 0 | 0 |
| 8 | 0 | 0 | 0 | 0 | 0 | 0 | 1 | 0 |

**Table 5.3(d): Cyclomatic Complexity from Connection Matrix for the Graph Matrix**

| | Connected to a Node | | | | | | | | Connection |
|---|---|---|---|---|---|---|---|---|---|
| **Node** | 1 | 2 | 3 | 4 | 5 | 6 | 7 | 8 | |
| 1 | 0 | 1 | 0 | 0 | 0 | 0 | 0 | 0 | 1-1=0 |
| 2 | 0 | 0 | 1 | 0 | 0 | 0 | 0 | 0 | 1-1=0 |
| 3 | 0 | 0 | 0 | 1 | 0 | 1 | 0 | 0 | 2-1=1 |
| 4 | 0 | 1 | 0 | 0 | 1 | 0 | 0 | 0 | 2-1=1 |
| 5 | 0 | 0 | 0 | 0 | 0 | 0 | 0 | 0 | --- |
| 6 | 0 | 0 | 0 | 0 | 0 | 0 | 1 | 1 | 2-1=1 |
| 7 | 0 | 0 | 0 | 1 | 0 | 0 | 0 | 0 | 1-1=0 |
| 8 | 0 | 0 | 0 | 0 | 0 | 0 | 1 | 0 | 1-1=0 |
| | | | | | | | | **Sum** | 3 |
| | | | | | | | | **Cyclomatic Complexity** | 3+1=4 |

Each letter has been replaced with a 1, indicating that a connection exists (zeros have been excluded for clarity). Represented in this form, the graph matrix is called a connection matrix. Table 5.3(d) depicts Cyclomatic Complexity calculation for the connected matrix. In the connection matrix, each row with two or more entries represents a predicate node. Therefore, performing the arithmetic shown to the right of the connection matrix in table 5.3(d) provides with a method for determining Cyclomatic Complexity.

## 5.4 BASIS PATH TESTING

Basis Path testing is a White-Box testing technique proposed by Tom McCabe. The Basis Path method enables to derive a logical complexity measure of a procedural design and use this measure as a guide for defining a basis set of execution paths. Test cases derived to exercise the basis set are guaranteed to execute every statement in the program at least once during testing. It is a hybrid of path testing and branch testing. Path testing is designed to execute all or selected paths through a computer program, while branch testing is designed to execute outcome of each decision point in that program.

Basis path testing tests all the independent paths that could be used to construct any arbitrary path, and fulfills the requirements for branch testing. A path is a sequence of instructions or statements that starts with an entry, junction, or decision and ends with another, or possibly the same, junction, decision, or exit. A path may go through several junctions, processes, or decisions, one or more times. An independent path is any path that introduces at least one new set of processing statements or a new condition.

### 5.4.1 Scope and Prospects

There may be many possible paths between the entry and exit of a software program. For decision statement the number of paths may be double, for case statements the number of path depends the number of cases, and in loop statement the number of paths increases with iteration values possible for the lop. Complete path coverage of even a simple unit is extremely difficult. Like other software development activities, software testing does have limited resources in terms of time, people, and equipment. The challenge is to select a set of test case that is most likely to identify as many defects as possible within the limited resources. There is a scope to reduce as much redundancy as possible from the testing process insuring adequate test coverage.

A basis path is a unique path where no iterations are allowed. Following steps may lead to generate test case in Basis Path Testing:

- Draw a control flow graph
- Calculate Cyclomatic complexity
- Choose a "basis set" of paths
- Generate test cases to exercise each path

### 5.4.2 Test Case Generation

Test cases are derived to exercise the basis set and to guarantee execution of every statement in the program at least once. Cyclomatic complexity V (G) is used to determine the number of tests to perform. V (G) will provide an upper bound of tests that must be executed to guarantee coverage of all statements. To derive a test case, following steps are recommended:

(a) Draw a corresponding flow graph using the design/code as foundation

(b) Determine the Cyclomatic Complexity of the resultant flow graph

(c) Determine a basis set of linearly independent paths

(d) Prepare test cases that will force execution of each path in the basis set

Figure 5.4.2(a) shows code and corresponding flow graph. Figure 5.4.2(b) shows an implementation to derive test case for the flow graph.

```
S0    public    void f ( ) {

S1    If (C>10) {

S2                   F=F+C;
                     Flag=1;

S3     } else if(C>5) {

S4                   F=F-1;
       } else {

S5                   F=C*12;

S6     }
```

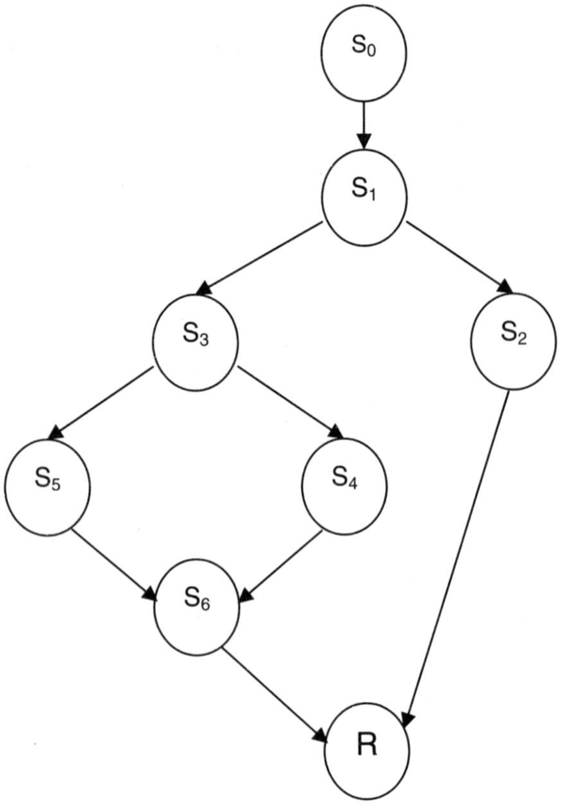


**Figure 5.4.2(a): Flow Graph for Code**

1. Flow graph for the code is shown in figure 5.4.2(a).
2. Cyclomatic complexity for the graph is 3 (e=9, n=8,p=1)
3. A basis set of linearly independent paths are

   Path-1: $S_0, S_1, S_2, R$

   Path-2: $S_0, S_1, S_3, S_5, S_6, R$

   Path-3: $S_0, S_1, S_3, S_4, S_6, R$

4. Test cases for

   Path-1: C= 12

   Path-2: C= 7

   Path-3: C= 3

**Figure 5.4.2(b): Test Case Generation for the Flow Graph**

## 5.5 CONTROL STRUCTURE TESTING

No doubt, basis path testing is simple and highly effective. But it is not enough in itself. Its variations are control structure testing, which broadens testing coverage and improves the quality of White-Box testing. There are several ways to perform control structure coverage. Important coverage techniques are described briefly in the following section.

### 5.5.1 Scope and Prospects

Control structure testing controls over the order in which the instructions in program are executed. One of the major objectives of control structure testing includes the selection of test cases to meet various criteria for covering the code. In control structure testing test cases are used to exercise as much of the program as possible according to some criteria. If some part of the program or component is not exercised then there may well be undiscovered faults lurking there.

### 5.5.2 Test Case generation

Test cases are derived to exercise control over the order of the execution of the code. Test cases generally correlate more to decisions than to statements. A number of different metrics are used to determine how well exercised the code is. It's usually a good idea to start with the simplest metrics and move on to the more powerful ones later. In practice this means starting with statement coverage, and when a maximum coverage has been achieved, move on to branch coverage, then to condition and path coverage. Coverage-based testing works by choosing test cases according to well-defined coverage criteria. The more common coverage criteria are the following:

- *Statement Coverage or Node Coverage:* Every statement of the program should be exercised at least once.
- *Branch Coverage or Decision Coverage*: Every possible alternative in a branch (or, decision) of the program should be exercised at least once. For if statements this means that the branch must be made to take on the values true and false.
- *Condition Coverage:* Each condition in a branch is made to evaluate to true and false at least once.
- *Decision/Condition Coverage:* Each condition in a branch is made to evaluate to both true and false and each branch is made to evaluate to both true and false.

**Conditions Testing:** A representation of the flow of control within a subprogram is the basis of the White-Box testing techniques. The first improvement to it is to ensure that the Boolean controlling expressions are adequately tested. This improvement is done through a condition testing. Condition testing is a test case design approach that exercises the logical conditions contained in a program module.

This method of testing concentrates on testing each condition in a program. The purpose is to determine not only errors in the conditions of a program but also other errors in the program. A number of condition testing approaches have been identified in the literature. Some are described in the following section.

*Branch Coverage:* An important technique used in testing of software is structural testing, in which a particular type of program element is selected for coverage. One of the examples of structural testing is the *Path Testing*, which requires exercising a fixed sequence of statements, which constitutes a path from the beginning of the program to a termination point. Other examples of the structural testing are statement, branch, and data flow coverage.

These testing requires exercising a given program element using any path through that element. Branch coverage is an important criteria used during the structural testing of programs. In branch testing every possible branch is tested, regardless of the outcome. It is important to find faults and to give credible information about the state of the component, on which business decisions can be taken. Ideally an independent person or body, to the one that wrote the code is responsible for performing branch coverage testing. However, usually, the developers conduct it at the point where the code was written.

Metric for branch decision coverage is available to measure how much of the code has been executed. This metric is used to track the amount of testing conducted on one or more components using branch testing. As this type of testing seeks, ideally, to execute every branch or decision outcome in the code, it is an important guide as to the amount of testing done, and the amount left.

Branch coverage ensures that whenever a program can jump, it jumps to all possible destinations. Figure 5.5(a) shows an example of if statement.

```
if (x>0)
{
   print "a";
}
else
{
   print "b";
}
```

**Figure 5.5(a): An example of If-Statement**

Full coverage is only achieved here only if (x>0) is true on one occasion and false on another. Achieving full branch coverage will protect against errors in which some requirements are not met in a certain branch. Another example is shown in figure 5.5(b), which may fail if $x is false.

*Statement Coverage:* Statement coverage may be defined as the percentage of executable statements in a component that have been exercised by a test case suite. It is also known as line coverage, segment coverage and basic block coverage. This measure reports whether each executable statement is encountered. Statement coverage identifies which statements contained within conditional blocks were executed during simulation.

It provides similar information to branch coverage, but the importance of statement coverage is that it provides a measure of the importance of a module. Statement coverage is performed by personnel as close as possible to the code or testing, like test analyst or tester. It is an ongoing process throughout the lifecycle as components are identified, test analysis can begin, and the metric calculated. Following advantages have been observed in statement coverage.

- Faults are evenly distributed through code, therefore, the percentage of executable statements covered reflects the percentage of faults discovered.
- This measure is affected more by computational statements than by decisions.

```
if ($x)
{
    $h = { a => 1 }
}
else
{
    $h = 0;
}
print $h->{a};
```

**Figure 5.5(b): An example of If-Statement**

Statement coverage is the most basic form of code coverage. A statement is covered if it is executed. Statement testing is a test case design technique for a component in which test cases are designed to execute statements. Statement testing is a structural or White-Box technique, because it is conducted with reference to the code. Statement testing comes under dynamic analysis. When a program is developed, every statement is expected to be executed, and should therefore be run at least once. Two important measurements important at the time of statement coverage by the test suite include the total number of statements and the number of statement to be tested. Following disadvantages have been observed with regard to statement coverage based testing:

- It is insensitive to some control structures, and is completely insensitive to the logical operators (|| and &&).
- It does not report whether loops reach their termination condition.
- It cannot distinguish consecutive switch labels.
- It considers do-while loops with the same rank as non-branching statements.

- In statement coverage, test cases generally correlate more to decisions than to statements.
- It is difficult to minimize the number of test cases and still ensure the execution of all statements.

A set of test cases $T=\{t_1, t_2, t_3, \ldots\ldots t_k\}$ satisfies the statement coverage criterion if and only if for, in a flow graph, any node $v \in V$, there exists a test case $t_i$ and its corresponding execution path $P_i$, where $v \in P_i$ . Figure 5.5(c) shows an example.

```
read (x); read (y);

if x > 0 then

          write ("X Greater Than 0");

else

          write ("X Less Than OR Equal to 0");

end if;

if y > 0 then

          write ("Y Greater Than 0");

else

          write ("Y Less Than OR Equal to 0");

end if;
```

**Figure 5.5(c): Example Code for Statement Coverage**

Generally one test for each statement should be made analyzing which inputs are needed to cause the statement to be executed. Following test cases cover all statements shown in figure 5.5(c):

{<x = 2, y = 3>, <x = - 13, y = 51>, <x = 97, y = 17>, <x = - 1, y = - 1>}

While the minimal cases are as follows:

{<x = - 13, y = 51>, <x = 2, y = - 3>}

It may be quite difficult to achieve 100 percent statement coverage. There may be sections of code designed to deal with error conditions, or rarely occurring events such as a signal received during a certain section of code.

*Code Coverage:* Code coverage is a way of ensuring that the tests are actually testing code. Code coverage will tell how much of code are exercised by running the test. Generally code coverage tries to cover more of the testing problem space to prove the absence of faults, or at least the absence of a certain class of faults. There are number of criteria to determine the exercised code. Statement coverage is one of them, which tells whether the statement has been exercised in code or not. Code coverage is a White-Box testing methodology which requires knowledge and access to the code itself rather than simply using the interface provided. Code coverage is probably most useful during the module testing phase. Code coverage also has benefit during integration testing and probably at other times, depending on how and what are being tested.

**Data Flow Testing:** Data flow testing is a form of White-Box testing, which is based on data flow analysis. No doubt, dataflow testing has been shown in several empirical studies to be more effective at revealing error than other, structure-based testing criteria. Data flow testing is a testing technique based on the observation that values associated with variables can effect program execution. The data flow testing method chooses test paths of a program based on the locations of definitions and uses of variables in the program.

Data flow testing not only explores program control flows but also pays attention to how a variable is defined and used at different places along control flows. This could lead to more efficient and targeted test suites than pure control-flow-based test suites. Data flow testing shows a way to distinguish between the useful ones and the less useful ones among test cases generated from pure control-flow-based testing techniques and minimizes the number of required test cases without reducing the effectiveness of the test suite. Data flow testing strategies are helpful for choosing test paths of a program including nested if and loop statements.

Data flow testing of procedural programs is based on both the control flow and data flow possible within a program. The main principle of data flow testing is if there is a value assignment to a variable that is not used in some subsequent computation or predicate in the program during the program test, it cannot be assured that the steps leading up to the assignments were correct. Data flow testing involves tracking input variables through a program, following them as they are modified, until they are ultimately used to produce output values. Ideally the program tester would like to provide as many different test cases as are necessary to exercise all paths between variable assignments and variable uses.

Data flow testing mainly focuses on testing the value assignment of each variable in a program by executing sub-paths from the assignment or *definition* to some program points in which the variable is used. The use of a variable can be classified as either a computation use or a predicate use. If the value of a variable is used in computing a value for defining other variables or as an output value in an output statement, the use is called *computation use*, denoted as *c-use*. Otherwise, if the value of a variable is used to decide the result of a predicate statement for selecting execution paths, the use is called *predicate use*, denoted as *p-use*. A *def-use pair* of a variable v is an order pair (d, u) such that d is a statement having a definition of v and u is a statement having a use of v, or some memory location bound to v, that can be reached by d over some path in the program.

**Loop Testing:** Loop testing is a White-Box testing approach that concentrates on the validity of loop constructs. Loops form the basis of most algorithms implemented using software. Most of the times, loops are not considered when conducting testing. Several loops have been identified in literature including the following:

- Simple Loops
- Nested Loops
- Concatenate Loops
- Unstructured Loops

These four types of the loops are described in the following section:

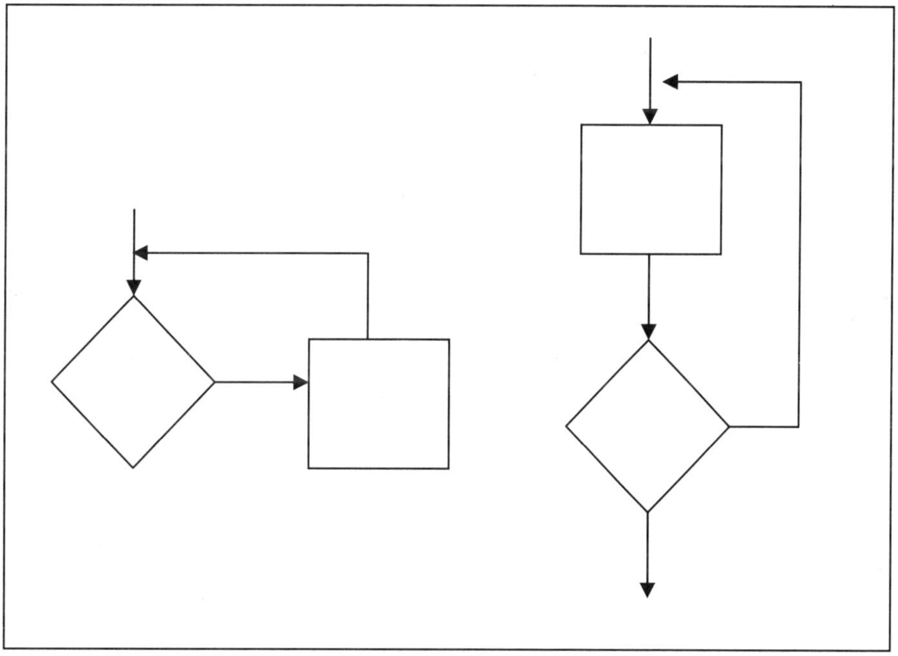

**Figure 5.5(d): Simple Loops**

*Simple loops:* Simple loops are depicted in figure 5.5(d). Several guidelines exists for executing test while using simple loops including following:

- Entirely skip the loop.
- Only one passes through the loop.

- Two passes through the loop.
- M passes through the loop where m<n.
- n-1, n, n+1 passes through the loop. (Where n is the maximum number of allowable passes through the loop)

*Nested loop:* The number of possible tests increases as the level of nesting grows in nested loop, and hence results in an impractical number of tests. Figure 5.5(e) depicts a nested loop. The number of tests may be limited with the following approaches:

- Begin with innermost loop. All other loops should be set to minimum values.
- Perform simple loop tests for the innermost loop by holding the outer loop at their minimum iteration parameter value.
- Work outward, performing tests for the next loop, but keeping all other outer loops at minimum values and other nested loops to 'typical' values.
- Conduct the test for all the loops (continue until all loops have been tested).

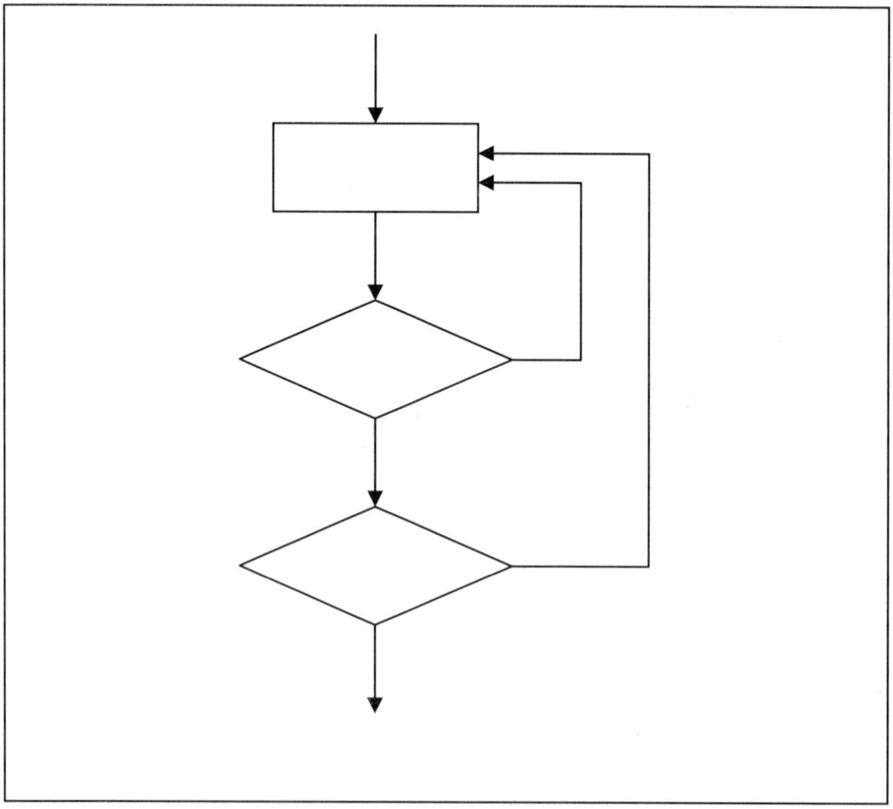

**Figure 5.5(e): Nested Loop**

*Concatenated Loop:* If each of the loops is independent of the other, concatenated loops can be tested using the techniques outlined for simple loops, otherwise the approach used for nested loops may be used to conduct the test for Concatenated loops. A concatenated loop is shown in figure 5.5(f).

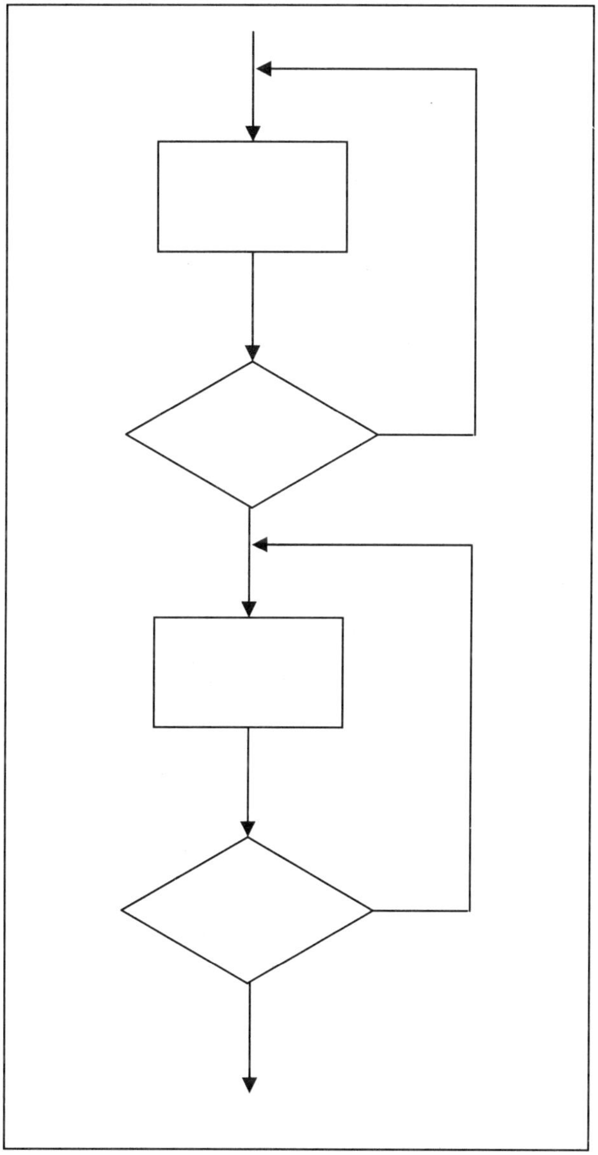

**Figure 5.5(f): Concatenated Loop**

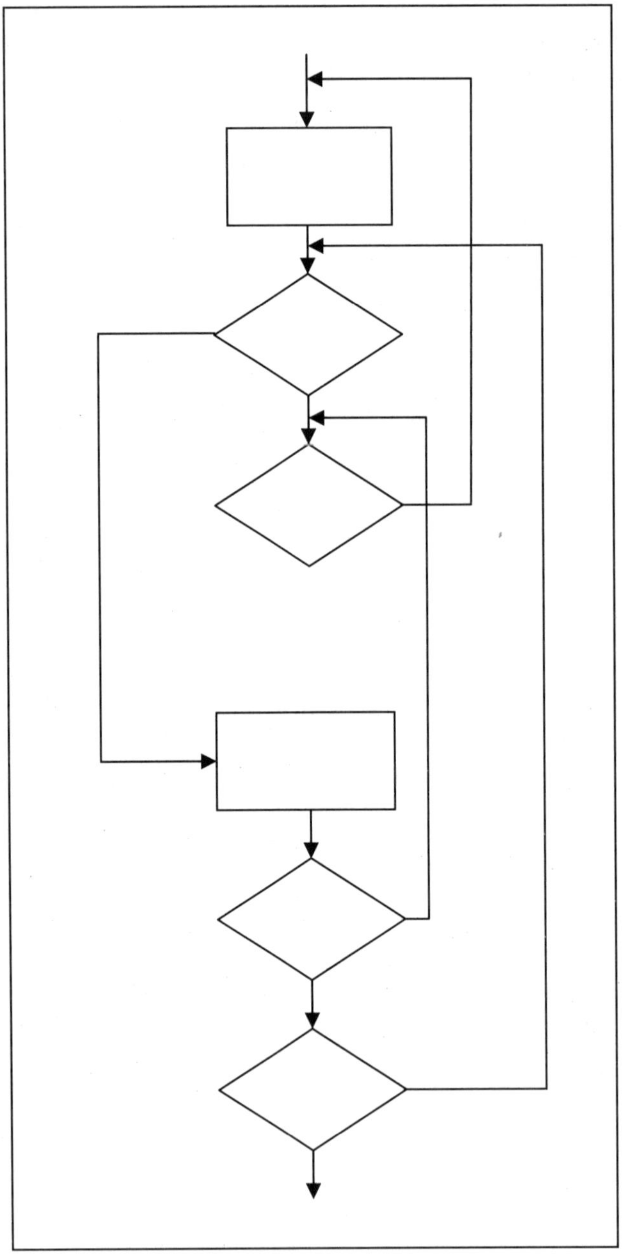

**Figure 5.5(g): Unstructured Loop**

*Unstructured Loop:* This class of loop is designed to reflect the use of the structured programming constructs. Unstructured loop is shown in figure 5.5(g).

## 5.6 MUTATION TESTING

Software testing is a vital yet expensive part of the software development process. It is evident from the studies and literature that it often consumes considerably high percentage of the total development budget. Thus approaches that automate or semi-automate sections of software testing may significantly improve the efficiency and quality of the software development process. Mutation testing is a fault-based technique used to test the quality of the test suite. Mutation testing is a technique for unit testing software that, although powerful, is computationally expensive. The principal cost of mutation is that many variants of the test program, called mutants, must be repeatedly executed. This cost has prevented mutation from becoming widely used in practical situations. But recent engineering advances have given several techniques and algorithms for significantly reducing the cost of mutation testing. Mutating certain statements in source code and checking if test code is able to find the errors do this.

### 5.6.1 Scope and Prospects

Mutation testing is one of the powerful techniques used to create more effective test suits to make the tester more confident about the program to be tested. The idea is very simple: given an appropriate set of mutation operators, if a test set kills the mutants generated by these operators then it is likely to be good at finding real faults. Following advantages of mutation testing have been observed by the experts and practitioners:

- It can uncover ambiguities in code.
- It leads to less buggy and more reliable product.
- It can be easily automated in comparison with some other forms of testing.
- By proper choice of mutant operations, comprehensive testing can be performed.
- It is at least as powerful as Data Flow testing

During mutation testing, many versions of a program are created. These created versions are mutated to seed a single fault. Now, the mutant program generated is run against the test cases. The goal is to get each faulty version fail. Each time a test case causes the faulty version to fail, the mutant is called killed mutant. The test case is saved. At last when all mutants have been killed, the saved test cases will comprise the test suite used to test the original program. Following disadvantages of mutation testing have been observed:

- Standard set of mutation operators may lead to a vast number of mutants.
- While a mutant is syntactically different from the initial program it may have the same behavior as the program. Clearly equivalent mutants can never be killed and thus it is important, but often difficult, to identify them.
- Some mutants, called stubborn mutants, may be non-equivalent but difficult to kill.

- Each mutant may need to be tested against many of the test cases before it can be distinguished from the original program.
- Mutation Testing is complicated and time-consuming to perform without an automated tool.
- Each mutation is equal in size and complexity to the original program, and a relatively large number of mutant programs need to be tested against the candidate test suite.

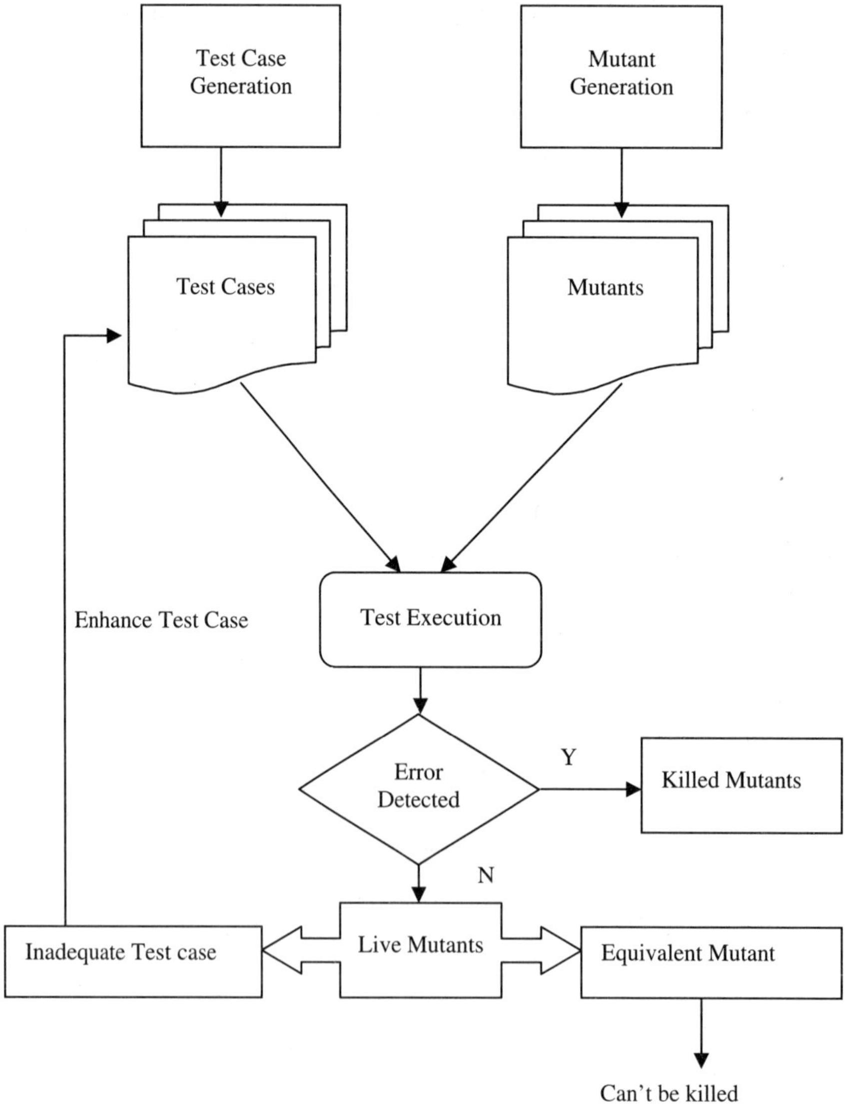


**Figure 5.6.2: Mutation Testing Process**

### 5.6.2 Test Case Generation

Mutation testing is a fault-based testing technique. This technique is based on the assumption that a program is well tested if all simple faults are predicted and removed; complex faults are coupled with simple faults and are thus detected by tests that detect simple faults. Mutation testing technique was first designed to create effective test data, with an important fault revealing power. It has been originally proposed by DeMillo in 1978, and consists of creating a set of faulty versions or mutants of a program with the ultimate goal of designing test cases that distinguishes the program from all its mutants. Figure 5.6.2 shows the entire process during mutation testing.

Mutation testing is the process of re-writing source code in order to flush out ambiguities that may exist in the code. These ambiguities can cause failures in software if not detected and fixed by the developer. Because these faults are often very subtle, they can easily pass conventional testing and debugging only to be passed on to the customer. The process of Mutation testing is based upon seeding the implementation with a fault. This is possible by applying a *mutation operator*, and determining whether testing identifies this fault. A mutated program is called a *mutant* and if a test case distinguishes between the mutant and the original program it is said to *kill* the mutant. Mutation testing may be used to judge the effectiveness of a test set: the test set should kill all the mutants.

## 5.7 GRAY-BOX TESTING

Over the past several years there has been a growing realization that software testing can no longer be treated as an informal art if software is to meet its requirements with high degree of reliability. Software testing is a critical software development activity used to detect errors in software. It is a very labor-intensive and expensive process. Debugging, along with the testing and maintenance, are crucial activities in software development processes. It has been revealed from the literature survey that the cost for correcting an error or a bug found after software release is four times more than fixing an error found at testing phase. The techniques for finding errors as early as possible in development processes are very significant to produce high quality software. Software testing is an important process of software quality assurance. Program testing is a significant process to assure and control software quality.

The Gray-Box Software Testing Methodology merges Black-Box, White-Box, assertion testing, and mutation testing methods into a seamless testing methodology. The gray box testing methodology is defined as the application of requirement based test cases generated from requirement specification language. It validates that the software meets its external specification and that all paths through the software have been exercised and verified. The Gray Box methodology is a full software life-cycle testing methodology used to functionally verify Ada, C, FORTRON and Assembly Language software developed for embedded application. This methodology reduces the cost and the time of software development by automatically creating all test unique software.

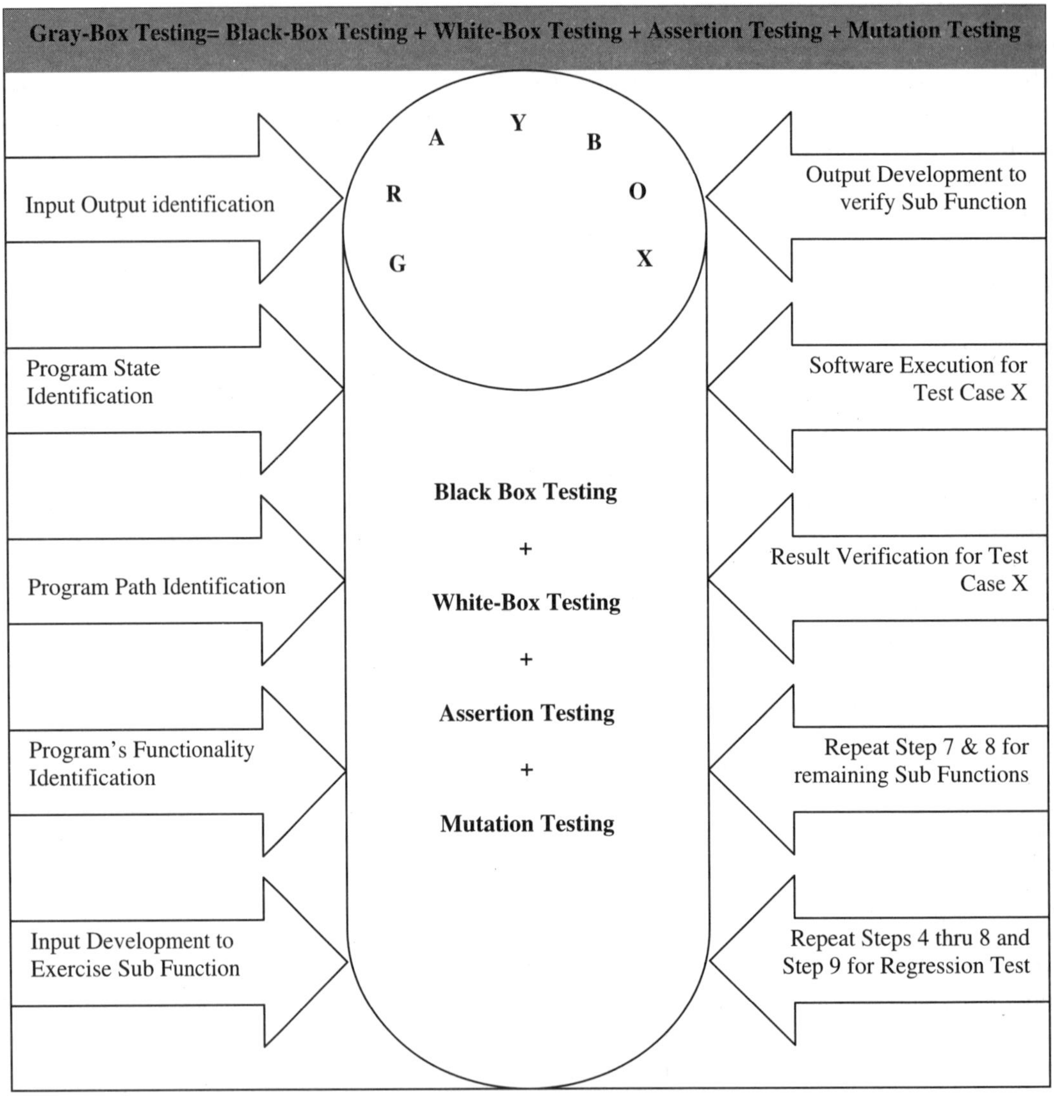


**Figure 5.7: Gray-Box Testing Methodology**

### 5.7.1 Scope and Perspectives

Various testing methodologies have been proposed in literature for testing software. Normally, software testers use *Black-Box* testing methodology while they are validating software requirements. Black-Box testing is the infusion of data into the software's interface and the extraction of its output at the software's interface. Similarly, for verifying software design, they use *White-Box* testing methodology. Software testers normally use a software debugger or emulator during White-Box testing to verify the software design. Another important type of testing methodology is *Assertion Testing*, which defines the validation of software by defining predicates and verifications. Predicates are the preconditions that must be valid before a program is executed. Predicates are verified to assure whether the correct output has been obtained. Assertion testing can be used with both Black-Box and White-Box methods of testing. Mutation testing is another special form of software testing that is destructive in nature, which calls for creating a variant of the software under test.

### 5.7.2 Test Case Generation

The Gray-Box methodology was developed especially to test embedded software systems. It is advocated as a successful methodology that will make formal verification an acceptable reality. Gray-Box testing is based on requirement based test case generation. It uses assertion methods to preset all the conditions required, prior to a program being tested. Formal verification is one of the commonly used techniques for ensuring that a core program is correct to a very large degree. If the requirement specification language is being used to specify the requirement, it would be easy to execute the requirement stated and verify it's correctness. Preconditioned predicates specified in assertion testing may be used in executing software requirements to generate the expected results, which is known as post-conditional verifications. The Gray-Box testing methodology will use the predicate and Verifications defined in requirement specification language as inputs to the requirements based test case generation phase. Gray-Box testing methodology is a ten-step process used to functionally validate embedded systems software. Figure 5.7 show steps involved in Gray-Box methodology.

## 5.8 CONCLUSION

In White-Box testing, software is viewed as a White-Box, or glass-box, since structure and flow of the software under test are visible to the tester. Glass box testing requires the intimate knowledge of program internals, while Black-Box testing is based solely on the knowledge of the system requirements. There are many techniques available in white-box testing, because the problem of intractability is eased by specific knowledge and attention on the structure of the software under test. Basis Path testing and Control Structure testing are two important types of testing technique available to conduct White-Box testing. Basis path testing tests all the independent paths that could be used to construct any arbitrary path, and fulfills the requirements for branch testing. Test cases are derived to exercise the basis set and to guarantee to execute every statement in the program at least one time during testing. Path coverage is a test case design technique for a component in which test cases are designed to execute branch outcomes is branch coverage testing, while statement coverage is the most basic form of code coverage. White-Box testing techniques rely on the structure of a program. Control-flow testing, loop testing, and data-flow testing, all maps the corresponding flow structure of the software into a directed graph. Test cases are carefully selected based on the criterion that all the nodes or paths are covered or traversed at least once. The boundary between black-box approach and white-box approach is not clear-cut. Many testing strategies mentioned, may not be safely classified into black-box testing or white-box testing. It is also true for transaction-flow testing, syntax testing, finite-state testing, and many other testing strategies. One reason is that all the techniques will need some knowledge of the specification of the software under test. Another reason is that the idea of specification itself is broad -- it may contain any requirement including the structure, programming language, and programming style as part of the specification content.

The first improvement to White-Box testing techniques is to ensure that the Boolean controlling expressions are adequately tested. Data flow testing shows a way to distinguish between the useful ones and the less useful ones among test cases generated from pure control-flow-based testing techniques and minimizes the number of required test cases without reducing the effectiveness of the test suite. Mutation testing is the process of re-writing source code in order to flush out ambiguities that may exist in the code. These ambiguities can cause failures in software if not detected and fixed by the developer.

## KEY TERMS

White-Box Testing, Clear Box, Glass-Box, Independent Path, Logical Decision, Path Completeness, Typographical Errors, Basis Path Testing, Control Flow Graph, Cyclomatic Complexity, Graph Matrix, Control Structure, Condition Coverage, Branch Coverage, Statement Coverage, Code Coverage, Data Flow Testing, Loop Testing, Object Oriented Software, Mutation Testing

## REVIEW QUESTIONS

## OBJECTIVE TYPE QUESTIONS

1. In White-Box testing, during test case design, all logical decisions should be executed_______.
   (a) On their true side only
   (b) On their false side only
   (c) On their true and false both the side
   (d) None of the above
2. In White-Box testing, during test case design, all loops should be executed_______.
   (a) At their boundaries
   (b) Within their boundaries
   (c) At their boundaries as well as within their boundaries
   (d) None of the above
3. In White-Box testing, path completeness is_______.
   (a) Insufficient

(b) Infeasible

(c) Feasible

(d) Sufficient

4. Basis path testing is a hybrid between_______.

(a) Path testing and branch testing

(b) Statement coverage and branch coverage

(c) Path testing and statement coverage

(d) Branch testing and branch coverage

5. White-Box testing techniques_______.

(a) Rely on the structure of program

(b) Do not rely on the structure of the program

(c) Rely on the execution of the program

(d) None of the above

6. Flow graph may be used to depict_______.

(a) Structure of program

(b) Logical flow of program

(c) Both structure and logical flow of the program

(d) None of the above

7. In a flow graph, a flow control is represented by_______.

(a) Lines only

(b) Arrows only

(c) Lines or arrows both

(d) None of the above

8. Condition testing determines_______.

(a) Error in Conditions

(b) Other errors in program

(c) Errors in conditions and other errors in program

(d) None of the above

## SHORT ANSWER TYPE QUESTIONS

1. What is White-Box testing?
2. White-Box testing should be called clear-box testing, why?

3. What do you understand by status of the program?
4. What do you understand by typographical errors? How can these errors be removed?
5. 'White-Box testing is insufficient'. Justify.
6. Why is it said that path completeness is usually infeasible in White-Box testing?
7. Describe the needs and importance of White-Box testing techniques.
8. Elaborate the significance of using White-Box testing techniques in software testing.
9. What do you understand by basis path testing?
10. Basis path testing said to be a hybrid between path testing and branch testing. Comment.
11. What do you mean by a path and an independent path? Explain with appropriate examples.
12. How can you draw a control flow graph?
13. What do you understand by Cyclomatic complexity? How can you calculate Cyclomatic complexity of a module?
14. How can a procedural design be translated into a control flow graph?
15. How many predicate nodes are required while testing compound Boolean expression using flow graph?
16. When is a graph said to be strongly connected?
17. How can you make a program control flow graph strongly connected?
18. How can Cyclomatic complexities indicate the effort required to test a component?
19. Describe the need and importance of graph matrix in software testing.
20. Though basis path testing is simple and highly effective, it is not enough in itself. Why? Justify.
21. What is condition testing? How it differs from branch coverage?
22. Define path coverage. Write down the path coverage criteria.
23. 'Branch coverage testing is a part of component testing'. Justify.
24. List out the metrics available for branch decision coverage.
25. How can branch coverage ensures that whenever a program can jump, it jumps to all possible destinations.
26. What do you understand by full coverage? How can you achieve it? What is the benefit of achieving full branch coverage?
27. '100% branch coverage implies 100% statement coverage'. Comment.
28. What is the importance of statement coverage in software testing?
29. Distinguish between statement coverage and branch coverage.

30. Statement coverage is said to be most basic form of code coverage. Comment
31. When is a statement said to be covered?
32. Why are statement coverages insensitive to some control statement?
33. What is loop testing? Explain with appropriate examples.

## DESCRIPTIVE QUESTIONS

1. In White-Box testing, how are test cases designed? Describe in detail. Present few more suggestions not included in this chapter for deriving test cases.
2. Logic errors and incorrect assumptions are inversely proportional to a path's execution probability. Justify.
3. Propose a suggestion/modification/amendment to make White-Box testing sufficient.
4. Why are iterations not allowed in basis path testing? Justify with the help of an example.
5. What is the basis set? How can you get a basis set of a path? Explain with the help of an example.
6. How can Cyclomatic complexity be used to pinpoint the areas of potential instability? Discuss in detail.
7. How can Cyclomatic complexity help in finding out the number of path to look for in basis path testing?
8. Though basis path testing is simple and highly effective, it is not enough in itself. Comment and Justify.

## REFERENCES

1. Sthamer, H. The Automatic Generation of Software Test Data Using Genetic Algorithms. PhD Thesis, University of Glamorgan, Pontyprid, Wales, Great Britain, 1996.
2. Baresel, A., Sthamer, H., and Schmidt, M. Fitness Function Design to improve Evolutionary Structural Testing. Proceedings of GECCO 2002, New York, 2002, 1329-1336.
3. Wegener, J., Baresel, A., and Sthamer, H. Evolutionary Test Environment for Automatic Structural Testing. Information and Software Technology, vol. 43, 2001, 841- 854.
4. Beizer, B. (1990). Software Testing Techniques. Boston, International Thompson Computer Press.
5. Beizer, B. (1995). Black-Box Testing. New York, John Wiley & Sons, Inc.

6. IEEE (1987). "ANSI/IEEE Standard 1008-1987, IEEE Standard for Software Unit Testing."
7. IEEE (1990). IEEE Standard 610.12-1990, IEEE Standard Glossary of Software Engineering Terminology.
8. Harman, M., Hu, L., Hierons, R., Baresel, A., and Sthamer, H. Improving Evolutionary Testing by Flag Removal. Proceedings of GECCO 2002, New York, 2002.
9. Harman, M., Hu, L., Hierons, R., Munro, M., Zhang, X., Dolado, J., Otero, M., and Wegener, J.
10. Kaner, C., J. Falk, et al. (1999). Testing Comptuer Software. New York, Wiley Computer Publishing.
11. McCabe, T. (1976). "A Software Complexity Measure." IEEE Transactions on Software Engineering **SE-2**: 308-320.
12. Pressman, R. (2001). Software Engineering: A Practitioner's Approach. Boston, McGraw Hill.
13. A Post-Placement Side-Effect Removal Algorithm. Proceedings der IEEE International Conference, New York, 2002.
14. D. Deveaux, P. Frison, and J.-M. Jezequel. Increase software trustability with self-testable classes in java. In Australian Software Engineering Conference (ASWEC), pages 3–11. IEEE Computer Society Press, 2001.
15. S. Beydeda and V. Gruhn. State of the art in testing components. In International Conference on Quality Software (QSIC), pages 146–153. IEEE Computer Society Press, 2003.
16. S. Beydeda, V. Gruhn, and M. Stachorski. A graphical representation of classes for integrated black- and white-box testing. In International Conference on Software Maintenance (ICSM), pages 706–715. IEEE Computer Society Press, 2001.
17. L. G. DeMichiel. Enterprise javabeans specification, version 2.1. Technical report, Sun Microsystems, 2002.
18. M. J. Harrold. Testing: A roadmap. In The Future of Software Engineering (special volume of the proceedings of the International Conference on Software Engineering (ICSE)), pages 63–72. ACM Press, 2000.
19. M. J. Harrold, D. Liang, and S. Sinha. An approach to analyzing and testing component-based systems. In International ICSE Workshop Testing Distributed Component- Based Systems, 1999.
20. M. J. Harrold and G. Rothermel. Performing dataflow testing on classes. In Second ACM SIGSOFT Symposium on the Foundations of Software Engineering (New Orleans, USA), volume 19 of ACM SIGSOFT Software Engineering Notes, pages 154–163. ACM Press, 1994.

21. H. S. Hong, Y. R. Kwon, and S. D. Cha. Testing of objectoriented programs based on finite state machines. In Second Asia-Pacific Software Engineering Conference (Brisbane, Australia), pages 234–241. IEEE Computer Society Press, 1995.
22. J. H¨ornstein and H. Edler. Test reuse in cbse using built-in tests. In Workshop on Component-based Software Engineering, Composing systems from components, 2002.
23. J.-M. Jezequel, D. Deveaux, and Y. L. Traon. Reliable objects: Lightweight testing for oo languages. IEEE Software, 18(4):76–83, 2001.
24. Orso, M. J. Harrold, and D. Rosenblum. Component metadata for software engineering tasks. In International Workshop on Engineering Distributed Objects (EDO), volume 1999 of LNCS, pages 129–144. Springer Verlag, 2000.
25. Y. L. Traon, D. Deveaux, and J.-M. Jezequel. Self-testable components: from pragmatic tests to design-to-testability methodology. In Technology of Object-Oriented Languages and Systems (TOOLS), pages 96–107. IEEE Computer Society Press, 1999.
26. Y. Wang, G. King, and H. Wickburg. A method for built-in tests in component-based software maintenance. In European Conference on Software Maintenance and Reengineering (CSMR), pages 186–189. IEEE Computer Society Press, 1999.
27. V. Aho, R. Sethi, and J. D. Ullman. Compilers, principles, techniques, and tools. Addison Wesley, 1988.
28. Atkinson and H.-G. Groß. Built-in contract testing in model-driven, component-based development. In ICSR Workshop on Component-Based Development Processes, 2002.
29. Baudry, V. L. Hanh, J.-M. Jezequel, and Y. L. Traon. Trustable components: Yet another mutation-based approach. In W. E. Wong, editor, Mutation testing for the new century, pages 47–54. Kluwer Academic Publishers, 2001.
30. S. Beydeda. The Self-Testing COTS Components (STECC) Method. ISBN 3-89975-462-X, Martin Meidenbauer Verlag, M¨unchen, 2004.
31. S. Beydeda and V. Gruhn. BINTEST - binary search-based test case generation. In Computer Software and Applications Conference (COMPSAC), pages 28–33. IEEE Computer Society Press, 2003.
32. S. Beydeda and V. Gruhn. Merging components and testing tools: The self-testing COTS components (STECC) strategy.
33. W. E. Howden. Symbolic testing and the dissect symbolic evaluation system. IEEE Transactions in Software Engineering, 4(4):266 – 278, Apr. 1977.
34. W. E. Howden and Y. Huang. Software trustability analysis. ACM Transactions on Software Engineering and Methodology, 4(1):36 – 64, Jan. 1995.
35. B.Korel. Automated software test data generation. IEEE Transactions on Software Engineering, 16(8):870 – 879, Aug. 1990.

36. B. Korel. Automated test data generation for programs with procedures. In S. J. Zeil, editor, International Symposium on Software Testing and Analysis, pages 209 – 215, San Diego, CA, Jan. 1996. ACM, ACM Press.

37. J. Laski and B. Korel. A data flow oriented program testing strategy. IEEE Transactions in Software Engineering, 9(5):33 – 43, May 1983.

38. Y. K. Malaiya. Antirandom testing: Getting the most out of black-box testing. Technical Report 96-129, Computer Science Department, Colorado State University, Fort Collins, CO, 1996.

39. B. Marick. General Coverage Tool (GCT), Jan. 1993. freeware.

40. G. Myers. The Art of Software Testing. John Wiley and Sons, New York, NY, 1979.

41. T. R. F. Nonweiler. Roots of low-order polynomial equations. Communications of the ACM, 11(4):269, Apr. 1968.

42. S. C. Ntafos. A comparison of some structural testing strategies. IEEE Transactions in Software Engineering, 14(6):868 – 874, June 1988.

43. T. J. Ostrand and M. J. Balcer. The category-partition method for specifying and generating functional tests. Communications of the ACM, 31(6):676 – 686, June 1988.

44. C. Ramamoorthy, S.Ho, and W. Chen. On the automated generation of program test data. IEEE Transactions on Software Engineering, 2(4):293 – 300, Apr. 1976.

45. J. M. Voas. A dynamic testing complexity metric. Software Quality Journal, 1(2):101 – 114, June 1992. Chapman and Hall.

46. J. M. Voas and K. W. Miller. The revealing power of a test case. Journal of Software Testing, Verification and Reliability, 2(1):25 – 42, May 1992. John Wiley and Sons.

47. J. M. Voas and K. W. Miller. Software testability: The new verification. IEEE Software, May 1995.

48. E. J. Weyukar. More experience with data flow testing. IEEE Transactions on Software Engineering, 19(9):912 – 919, Sept. 1993.

49. E. J.Weyukar and T. J. Ostrand. Theories of program testing and the application of revealing sub-domains. IEEE Transactions in Software Engineering, 6(3):236 – 246, May 1980.

50. L. J. White. Basic mathematical definitions and results. In B. Chandrasekaran and S. Radicchi, editors, Computer Program Testing, pages 367 – 375. North Holland, 1981.

51. W. E. Wong. On mutation and dataflow. PhD thesis, Computer Science Department, Purdue University, 1993.

52. H.Yin, Z. Lebne-Dengel, and Y.K.Malaiya. Automatic test data generation using checkpoint encoding and antirandom testing. Technical Report 97-116, Computer Science Department, Colorado State Univeristy, Fort Collins, CO, 1997.

53. H. Zhu, P. A. V. Hall, and J. H. R. May. Software unit test coverage and adequacy. ACM Computing Surveys, 29(4):365 – 427, Dec. 1997

## USEFUL LINKS

http://www.ddj.com/articles/2000/0003/0003a/0003a.htm

http://www.bullseye.com/coverage.html

http://www.eiffel.com/doc/manuals/technology/contract/page.html

http://www.mcgsoft.com/ftp/assurance.pdf

http://japan.amc.com/support/resources/white_papers/desdeb/desdeb.html

http://www.arl.wustl.edu/~lockwood/class/cs306/books/artofasm/Chapter_10/CH10-3.html

http://www.aqit.uwo.ca/□corless.

www.wiley.com/college/busin/icmis/oakman/outline/chap05/slides/control.htm

www.enterprisedb.com/documentation/spl-control-structures.html

www.pse.uri.edu/evs536/asp_book_notes_Word/ch05.doc

www.cgmopen.org/technical/testing/task2-report-TOC.html

documents.wolfram.com/v5/TheMathematicaBook/PrinciplesOfMathematica/EvaluationOfExpressions/2.6.9.html

www.csc.fi/cschelp/kaytto/script.html.en

withglee.com/Bases/GBCSGen.htm

withglee.com/Comments/GCCSIf.htm

www.scheme.com/csug/control.html

www.strath.ac.uk/CC/Courses/fortran/control1.html

www.loria.fr/~zendra/publications/jvm-2002.pdf

www.psoug.org/reference/control_struct.html

## Chapter 6

# Software Testing Strategies

*You've got to be careful if you don't know where you're going 'cause you might not get there!*
**Yogi Berra**

## OBJECTIVES

A successful completion of this chapter will enable the reader for the following:

- To understand and describe strategy in general.
- To appreciate the need, importance and significance of software testing strategy.
- To identify software testing strategic issues.
- To define and understand the strategic approach to software testing.
- To realize the facts related to verification and validation processes.
- To enumerate and realize generic view of software testing strategy.
- To identify techniques for Unit testing.
- To elaborate Integration testing, its approach and inputs and outputs.
- To finalize the techniques for Integration testing.
- To distinguish between Top-Down and Bottom-Up approach of Integration.
- To describe Validation testing, its advantages and disadvantages.
- To distinguish between Alpha and Beta testing.
- To identify techniques for System testing.
- To realize the need and importance of Stress testing.
- To identify the techniques for regression testing.
- To find out the condition for completion of testing.
- To chose software testing model.
- To identify the criteria for completion testing.
- To formulate a customized strategy for testing a software..

## 6.1 INTRODUCTION

A strategy outlines what to plan, and how to plan it. No doubt, a successful strategy is the guide through change. It provides a firm foundation for optimal and effective implementation of testing techniques. Unlike a plan, which is obsolete from the point of creation, a strategy reflects the values of an organization and remains current and useful. A good strategy stands as a clear counter to reactive, counter-productive test approaches. At the time of testing its products or tools, an organization tries to compare them against its expectations and values. By its nature, testing introduces change as problems are identified and resolved. A test strategy is necessary to allow these two impulses to work together. Furthermore, testing can never be said to be complete, and a core skill in testing is the justified management of conflicting demands; without a strategy, these judgments will be inconsistent to the point of failure. Any testing strategy needs to include test planning, test case design, test execution, and the resultant data collection evaluation. A test strategy indicates how testing is to be effectively and efficiently carried out. In order to make the best possible use of resources and time, it is decided on which parts and aspects of the system the emphasis should fall. The test strategy forms an important basis for a structured approach to testing and makes a major contribution to a manageable test process.

A testing strategy may be based on risk assessment. A risk is the chance of an error occurrence related to the damage expected when this error does occur. The developer of test strategy should realize that users are the best people to assess the damage and the frequency of use when valuing the risks. On the other hand, the members of software development team are best to assess the chance of error. During development of testing strategy, the developers should ensure that the test will be organized to find out most important problems in early stages of software development life cycle at the lowest costs. Another important aim in developing testing strategy is to find out the problems that require the most rework time first. The stepwise definition of testing strategy can be used for any test level.

Software development is a creative process. A strategy for software testing integrates software test case design techniques into a well-planned set of steps that cause the effective and efficient testing of software. It provides a road map for the software developer, the quality assurance organization, and the customer. A testing strategy is a vital enabler to this process-keeping focus on core values and consistent decision-making to achieve desired goals with the best use of resources. A software testing strategy should be flexible enough to promote the creativity and customization that are required to adequately test all large software-based systems.

A strategy for software testing may also be viewed in the context of spiral. *Unit testing* begins at the detailed design phase, and concentrates on each unit, that is, component of the software as implemented in source code. It proceeds to *Integration Testing*, where the focus is on design and the construction of the software architecture. At the *System Testing,* the software and other system elements are tested as a whole. Finally, *Acceptance Testing* addresses the testing of the system documentation by the user representative. It corresponds to the requirements phase of the software development lifecycle.

## 6.2 STRATEGIC ISSUES

A strategy for testing must include low-level tests that are required to verify that a small source code segment has been implemented correctly as well as high-level tests that validate major system functions based on customer requirements. A range of issues related to development of an appropriate strategy exists till now and debated regularly aiming the fraternity. A selection from those is described briefly as follows:

**Quantifiable Specification of Requirement:** One of the major objectives of testing is to uncover maximum number of errors. A good testing strategy also assesses other quality characteristics including portability, maintainability, flexibility, and usability. These quality characteristics should be well specified in a way that is measurable to avoid unambiguous testing result.

**Precise Statement of Testing Goals and Objectives:** The goals and objectives of software testing should be stated precisely and in measurable terms in order to make the testing process successful. The same should be stated within the test plan.

**Assuring User's Acceptability:** It is to understand the users of software and to develop a profile for each user category. Use-cases may be used to describe the interaction scenario for each class of users. These use-cases may be used to reduce overall testing effort by focusing testing on actual use of product.

**Designing Automated Testing:** Software should be build as robust as it can be designed to test itself. Such software should be capable of diagnosing certain classes of errors, and should accommodate automated testing and regression testing.

**Monitoring and Improving Testing Process:** A continuous improvement approach should be developed to monitor and enhance the effectiveness of software testing process. Testing strategy should be measured and monitored. Software testing metrics may be used to control the software testing process.

## 6.3 STRATEGIC PREMISES

Testing is a well organized set of activities, planned in advance and performed systematically. Therefore, there is a need to develop a set of stages where particular tests case design techniques and test approaches can be placed for the software engineering procedure. Various testing strategies have been identified in the literature, which provide a template for testing. Generally, all of these strategies have the following basic assumptions embedded in them in one way or the other:

- Software testing should better start at the modular level and work outward towards the integration of the complete system.
- Testing should be largely considered as the inductive process.
- A bottom-up approach would be more useful for testing.
- Diverse testing techniques would prove to be appropriate at diverse points in time.

- Software developer should conduct testing but they must be assisted by independent test groups in the larger interest.
- Independent testers remove the inherent conflict of interest, developers get his or her own product right.
- Testing and debugging are two diverse activities, but debugging is integral to any testing strategy.
- Testing should begin at the component level and work outward towards the integration of the entire computer-based system.
- Verification issues should focus on 'Are we building the product right?'.
- Validation issues should focus on 'Are we building the right product?'.

A software testing strategy must accommodate low-level tests that verify whether a small source code segment has been correctly implemented. It also checks high-level tests to validate if major system functions against customer requirements. A strategy must provide guideline for the practitioners and a set of milestones for the managers.

However, one of the important objectives of testing is verification and validation, being performed through different means. Verification is nothing but the set of operations that ensures whether the software correctly implements a particular function. It involves reviews and meetings to evaluate documents, plans, code, requirements, and specifications. This can be done with checklists, issues lists, walkthroughs, and inspection meetings. While validation is related to a different set of activities that ensures whether the software that has been produced is traceable to customer needs or not and involves actual testing and takes place after verifications are completed.

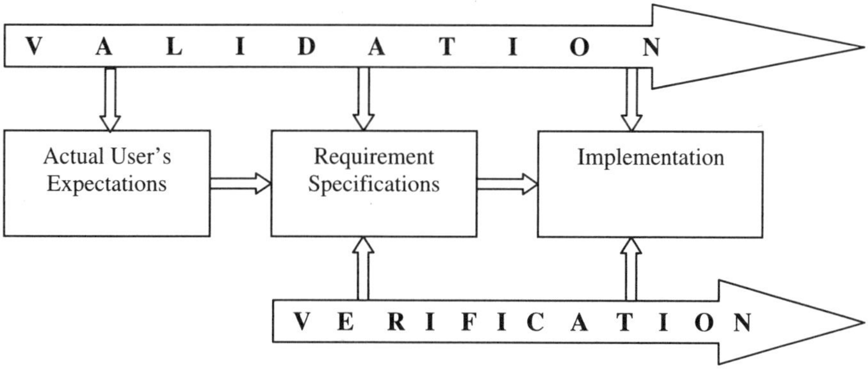


**Figure 6.3: Verification Vs Validation**

Verification activities aim at checking the correspondence of an implementation with its specification. While the activities that aim at checking the correspondence between a system and users' expectations are called *validation* activities. The distinction between validation and verification is informally illustrated in figure 6.3.Validation and verification activities complement each other. Verification can involve all development stages with users' reviews of requirements and design specifications. The main validation activities concentrate on the final product that can be extensively tested by the users during acceptance testing.

Since users' needs and the late execution of validation activities lead to high costs and risks, therefore validation must be paired along with the verification activities, which can be extensively executed at early stages of software development life cycle. Verification and validation of different properties require specific approaches. Dependability properties can be verified using functional and model-based testing techniques while usability properties require special-purpose techniques for their validation. However, it is notable that (i) how a practitioner decides on the best course of action during validation and verification; (ii) what necessary tools or techniques are being used for validation and verification purpose; and (iii) what is the best approach?

## 6.4 A GENERIC TESTING STRATEGY

Generally, people who produce the software are required to test it. Unfortunately, these developers have the major interest in showing that the program is error free. They try to show that the software developed matches with the customer's needs and was completed on time and within budget. An independent test group (ITG) works with the developer through out the project to ensure that the testing is carried out at the correct level. The software engineering procedures can be seen as spiral activities. Initially the systems engineering states the role of the software and lead the software requirement analysis, where the information domain, function, behavior, performance and validation criteria for the software are identified. Moving inwards along the spiral, we come to design and finally coding.

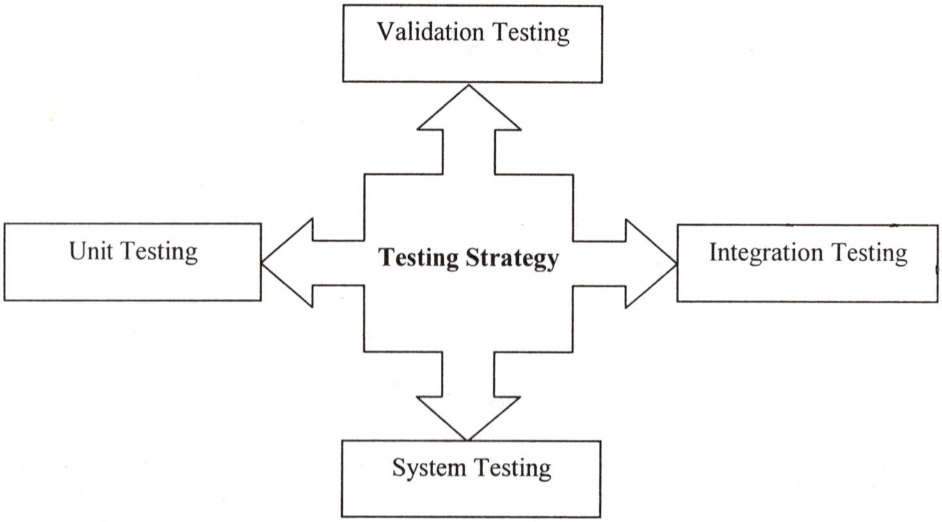


**Figure 6.4: Software Testing Strategy: A Generic View**

A strategy for software testing may be to move upward along the spiral. Unit testing happens at the vertex of the spiral and concentrates on each unit of the software as implemented by the source code. Testing happens upwards along the spiral to integration testing, where the focus is on design and the production of the software architecture. Finally system testing is performed, where software and other system elements are tested together. Figure 6.4 depicts a generic view of software testing strategy.

### 6.4.1 Unit Testing

Unit testing concentrates verification on the smallest element of the program – the module. One of the goals of the unit testing is to evaluate some piece, for example, file, program, module, component, etc. in isolation. Unit testing is the testing of individual components/units of the software. It is usually conducted as part of a combined code and unit test phase of the software lifecycle, although it is not uncommon for coding and unit testing to be conducted as two distinct phases. The unit test is the lowest level of testing performed during software development, where individual units of software are tested in isolation from other parts of a program.

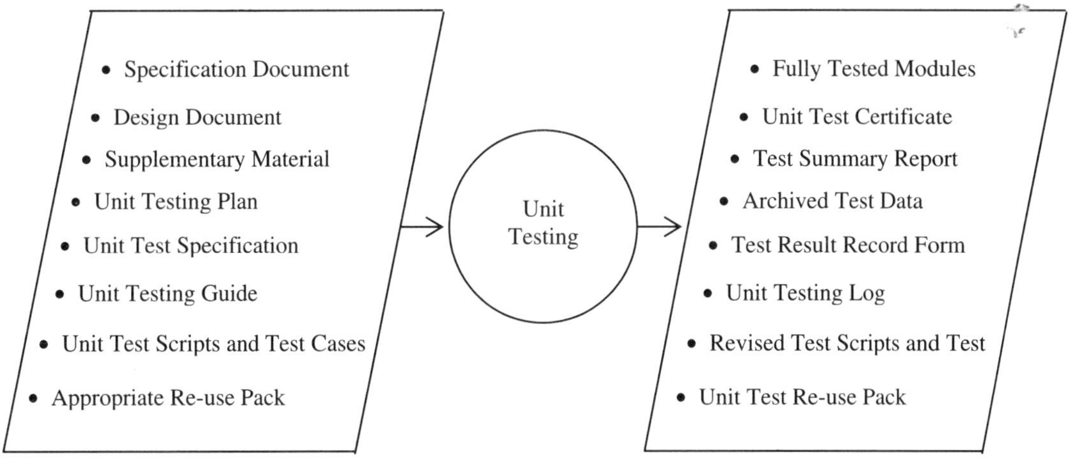


**Figure 6.4.1: Unit Testing: Inputs-Outputs**

Important control paths are tested to establish errors within the bounds of the module using the detailed description. Unit level testing is not just intended for one-off development use, to aid bug free coding. Unit tests have to be repeated whenever software is modified or used in a new environment. Consequently, all tests have to be maintained throughout the life of a software system. Other activities, which are often associated with unit testing, are code reviews, static analysis and dynamic analysis. Static analysis investigates the textual source of software, looking for problems and gathering metrics without actually compiling or executing it. Dynamic analysis looks at the behavior of software while it is executing, to provide information such as execution traces, timing profiles, and test coverage information. The testing techniques, which are generally considered appropriate for Unit testing include functional testing, structural testing, error-based testing, static testing, state transition testing and non-functional testing. Some of the important one are described in table 6.4.1 as follows.

**Table 6.4.1: Unit Testing Techniques**

| S.N. | Class of Testing | Testing Technique | Remarks |
|---|---|---|---|
| I | Functional Technique | Input Domain Testing | Design test cases that are representative of the range of allowable input, including high, low, and average values. |
| | | Equivalence Partitioning | Partition the range of allowable input so that the program is expected to behave similarly for all inputs in a given partition, then pick a test case from each partition. |
| | | Boundary Value | Choose test cases with input values at the boundary (both inside and outside) of the allowable range. |
| | | Syntax Checking | Choose test cases that violate the format rules for input. |
| | | Special Values | Design test cases that use input values that represent special situations. |
| | | Output Domain Testing | Design test cases that will produce output at the extremes of the output domain. |
| II | Structural Technique | Statement Testing | Ensure the set of test cases exercises every statement at least once. |
| | | Branch Testing | Design test cases so that each branch of if/then statement is exercised. |
| | | Conditional Testing | Design test cases so that each truth statement is exercised both true and false. |
| | | Expression Testing | Design test cases so that every part of every expression is exercised. |
| | | Path Testing | Design test cases so that every path is exercised (impossible in practice). |
| III | Error-Based Techniques | Fault Seeding | Create mutants of the program by making single changes, then run test cases until all mutants have been killed, one by one. |
| | | Mutation Testing | Create mutants of the program by making single changes, then run test cases until all mutants have been killed. |
| | | Historical Test Data | An organization keeps records of the average numbers of defects in the products it produces, and then tests a new product until the number of defects found approaches the expected number. |

The planning of Unit testing is the responsibility of the development team leader and should be developed with reference to the overall development and testing plan for application under test. It is assumed that Unit testing will not take place in the live environment but in the development environment or in a separate testing environment. The time and duration of Unit testing must be integrated into the development plan for the application under test in order to minimize the impact on the development process caused by Unit testing. Figure 6.4.1 depicts the inputs to Unit testing before test execution begins, and the generated outputs from Unit testing.

By convention, a Unit is allocated to a programmer for programming. He has to use functional specifications document as input to proceed. He prepares program specifications for his unit from the functional specifications. Program specification describes the programming approach and coding tips for the Unit's coding. Using these program specifications as input, programmer prepares unit test case documents for that particular unit. A unit test cases checklist may be used to check the completeness of unit test cases document. Program specifications and unit test cases may be reviewed and approved by quality assurance analyst, or by peer programmer. The programmer implements some functionality for the system to be developed. The same is tested by referring the unit test cases. While testing that functionality if any defects have been found, they are recorded using the defect logging tool whichever is applicable. The programmer fixes the bugs found and tests the same for any errors. Table 6.4.1 depicts the unit testing techniques, appropriate to be employed for unit testing.

### 6.4.2 Integration Testing

After testing all the individual units there is a need to test how they were put together to ensure no data is lost across interface, one module does not have an adverse impact on another and a function is not performed correctly. Integration testing is a systematic approach that produces the program structure while at the same time producing tests to identify errors associated with interfacing. Integration testing deals with many communicating modules. Big bang testing, which waits until all modules are integrated, is rarely effective, since integration faults may hide across different modules and remain uncaught. Most integration testing strategies suggest testing integrated modules incrementally. Integration strategies can be classified as structural and feature-driven. Structural strategies define the order of integration according to the design structure and include bottom-up and top-down approaches, and their combination.

The objective of integration testing is to take unit tested components and build a program structure that has been dictated by design. The quality of single modules is necessary but not sufficient enough to guarantee the quality of the overall system. The failure of low quality modules fatally leads to system failures that are often difficult to diagnose, and hard and expensive to remove. Unfortunately, many subtle failures are caused by unexpected interactions among well-designed modules. Following testing techniques are appropriate for Integration testing:

- *Functional Testing* using Black Box testing techniques against the interfacing requirements for the module under test.

- *Nonfunctional Testing* for performance or reliability testing of the module

The planning of Integration testing is the responsibility of the development team leader and should be developed with reference to the overall development and testing plan for application under test. It is assumed that Integration testing will not take place in the live environment but in the development environment or in a separate testing environment. The time and duration of Integration testing must be integrated into the development plan for the application under test in order to minimize the impact on the development process caused by Unit testing. Figure 6.4.2(a) depicts the inputs to Integration testing before test execution begins, and the generated outputs from Integration testing.

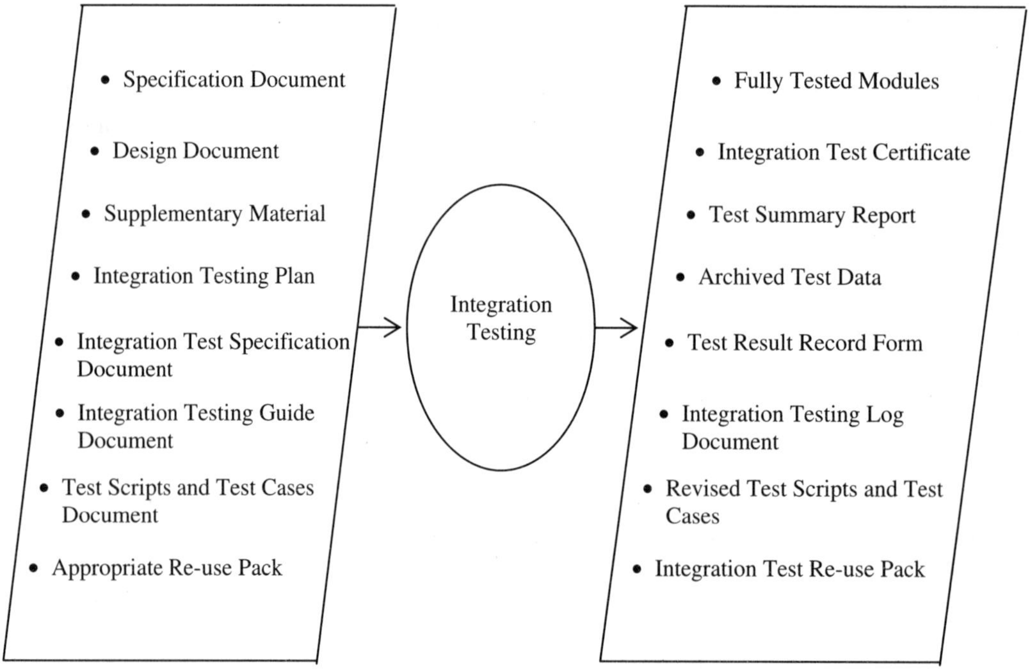


**Figure 6.4.2(a): Integration Testing: Input/Output**

Integration faults are ultimately caused by incomplete specifications or faulty implementations of interfaces, resource usage, or required properties. Unfortunately, it may be difficult or cost-ineffective to specify all module interactions completely. Integration faults can come from many causes:

- Inconsistent interpretation of parameters or values;
- Violations of value domains or of capacity/size;
- Side effects on parameters or resources, when modules use resources not explicitly mentioned in their interfaces;

- Missing or misunderstood functionality, when incomplete specifications are badly interpreted;
- Non-functional problems, which derive from under-specified non-functional properties, like performances; and
- Dynamic mismatches, which derive from unexpected dynamic bindings.

Figure 6.4.2 (b) shows the hierarchy of the modules. Depth-first integration would integrate these modules on a major control path structure, which is arbitrary and relies on application particular features. Modules 1, 2, and 5 may be integrated in one of the integration possibilities. Then there is option to integrate either module 8 or 6. Then after central or right hand control path are produced. While in case of breath-first integration, all the modules directly subordinate at each level will be integrate by moving across the structure horizontally. That is modules 2, 3, and 4 may be integrated first, and then modules 5 and 6. However, most commonly followed strategies for integration testing are Top-down Integration and Bottom-up Integration., described briefly as follows.

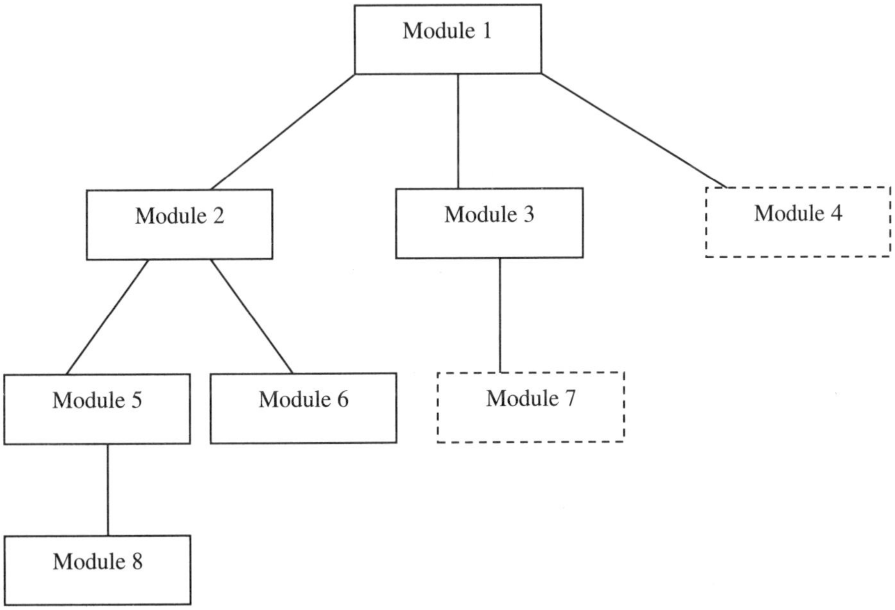


**Figure 6.4.2(b): Hierarchy of the modules**

**Top-Down Integration:** Top-Down integration is an incremental approach to gradually build the program structure. Modules are integrated by moving downward through the control hierarchy, beginning with the main control module. Modules subordinate to the main control module are incorporated into the structure using either a depth-first or breadth-first approach. Figure 6.4.2(c) gives a pictorial representation of top down integration approach.

The Integration process is performed in a series of following steps:

1. The main control module is identified as a test driver.
2. Stubs are substituted for all components directly subordinate to the identified test driver.
3. Selected subordinate stubs are replaced one at a time with actual components based on the integration approach.
4. Tests conducted as each component is integrated at last.
5. Finally, on completion of each set of tests, stub is replaced with the real component.
6. To ensure that new errors have not been introduced, regression testing may be conducted at the end.

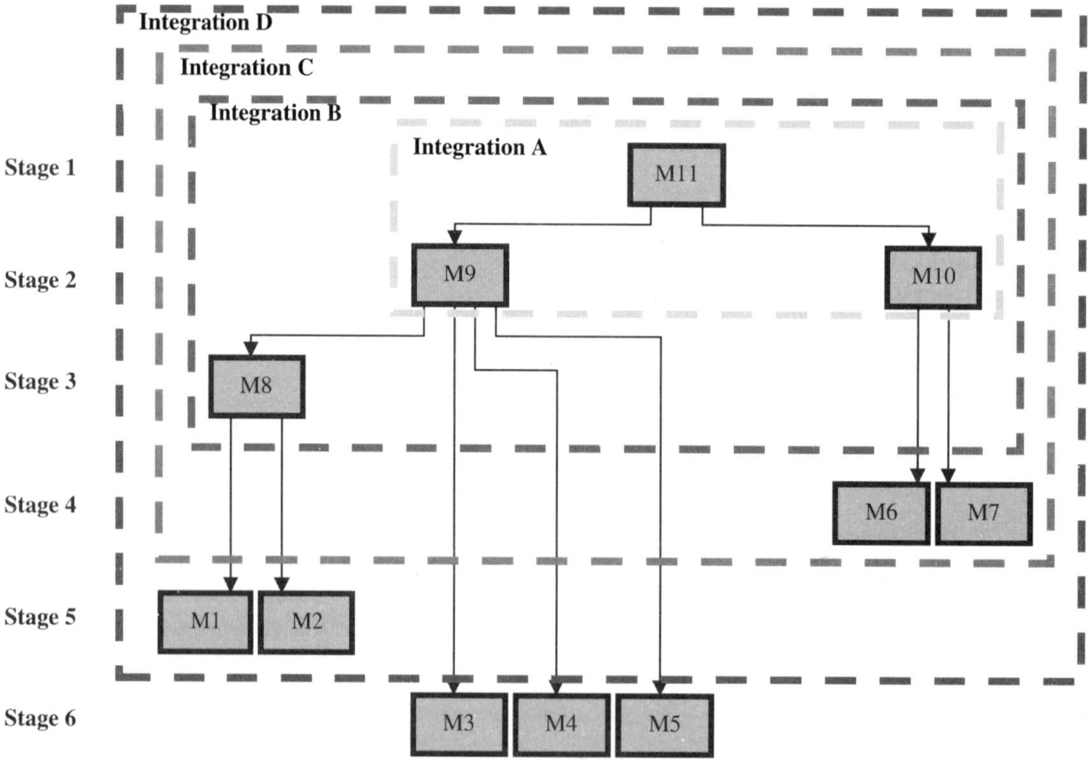


**Figure 6.4.2(c): Top-Down Integration**

In top down integration, application skeleton exists early in the software development life cycle and hence opens the opportunity for demonstration. User interface is done first in top down integration. Another advantage of using top down integration is that the test cases derived from requirements may be performed in multiple cycles of the process. On the other hand, top down integration is very expensive as writing stubs for untested modules are very costly. Another disadvantage of Top-Down integration is that the errors can be made in the stubs, resulting in false negatives or false positives.

**Bottom-Up Integration:** Bottom-Up integration testing begins with testing the modules at the lowest level, which are said to be atomic modules. As modules are integrated bottom up, processing required for modules subordinates to a given level is always available and the need for stubs is eliminated. Figure 6.4.2(d) gives a pictorial representation of Bottom-Up integration approach. A Bottom-Up integration strategy may be implemented with the following steps:

1. Low-level modules are combined into clusters.
2. Combined cluster usually perform a particular software sub function.
3. The cluster is tested.
4. A driver is written to coordinate test cases input and output.
5. Drivers are removed and clusters are combined moving upward in the program structure.

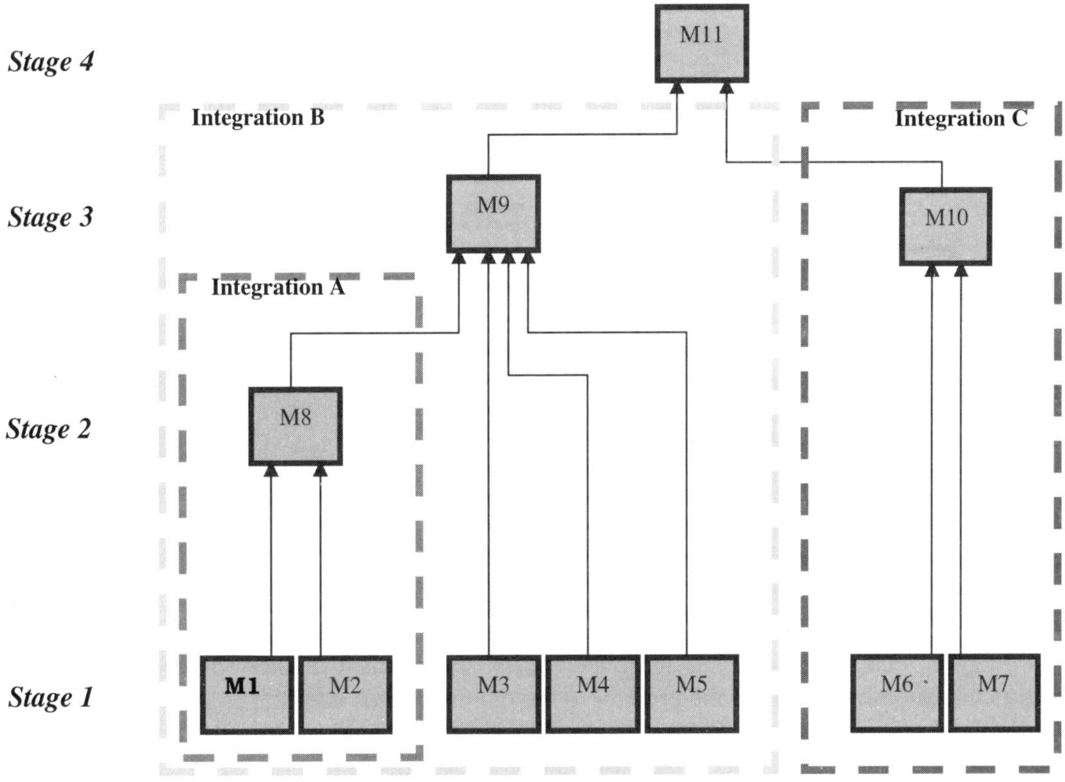


**Figure 6.4.2(d): Bottom Up Integration**

Creating test drivers and swapping them for higher-level components may aid in discovering false assumptions made by that component of a lower component. The major disadvantage of top-down approach is the need for stubs and the difficulties that are linked with them. Problems linked with stubs may be offset by the advantage of testing major control functions early. Another drawback of bottom-up integration is that the program does not exist until the last module is included.

### 6.4.3 Validation Testing

Validation can be defined in different ways. In general software is called valid if it functions in a fashion as expected by the customer(s). As a culmination of testing, software is completely assembled as a package. Interfacing errors are identified and corrected. As a final set of software tests, validation testing is started and is achieved through a series of black box tests that show conformity and consistency with requirements. One of the important elements of the validation process is a configuration review, which ensures that all the components of the software configuration have been properly developed.

Validation tests measure some aspect of an implementation's behavior against an expectation and answer whether the software complies with the expectation. For example, a test that measures the time required to perform a specific task is not a validation test. A test that measures the time required to perform a specific task and compares it to a benchmark is a validation test. Validation tests provide assurance that the software as written meets requirements and/or specifications. When validation testing becomes a fundamental part of the software development process, software is more likely to be completed, more likely to be fit for the intended purpose, is less expensive to build and maintain, and takes less time to complete.

The process of devising validation tests for a software module is quite difficult. This is because a prerequisite for devising validation tests is a detailed and specific understanding of the desired behavior of the module. At the outset of the development process, the intended behavior of the modules is not well understood. Many details are missing. The act of devising validation tests drives out uncertainty and forces the development team to make decisions. In my experience, writing validation tests is the most important part of the software design process. After each validation test case has been conducted, the function or performance characteristic conforms to specification and is accepted. During validation testing a deviation from specification is uncovered and a deficiency list is prepared. These deviations or errors discovered during validation test may rarely be corrected prior to scheduled delivery. Two general and typical validation testing are popularly known as Alpha and Beta testing.

**Alpha Testing**: It is very difficult to develop to determine how the customer will actually use the program. Therefore, a set of acceptance tests is performed to allow the user to check all requirements before shipping the product to the end user. This acceptance test can range from an informal test drive to rigorous set of tests. It may be conducted by the end user instead of the developer. Most developers use alpha and beta testing to identify errors that only users seem to be able to find. Its objective is to identify any serious errors and to judge if the indented functionalities are implemented. In a software development cycle, depending on the functionalities the number of alpha phases required is laid down in the project plan itself. Alpha testing is performed at the developer's sites, with the developer checking over the customers shoulder as they use the system to determine errors. During this phase, the test plan and test cases for the beta testing is created.

**Beta Testing:** Beta testing is conducted at more than one customer locations with the developer not being present. The beta test is a live operation of the software in an environment that cannot be controlled by the developer. Major objective of beta testing is to evaluate software technical contents and any hidden errors and side effects. It evaluates software ease of use and user documentation draft, and identifies and reports errors. The Software reaches beta stage when most of the functionalities are operational. The software is tested in customer's environment, giving user the opportunity to exercise the software, find the errors so that they could be fixed before product release. Beta testing is a detailed testing and needs to cover all the functionalities of the product and also the dependent functionality testing. Hence it is essential that it is planned well to get the task accomplished. The test plan document has to be prepared before the testing phase is started, which clearly lays down the objectives, scope of test, tasks to be performed and the test matrix which depicts the schedule of testing. The customer reports any problems they have to allow the developer to modify the system. The errors reported are documented internally for the testers and developers reference.

### 6.4.4 System Testing

Each software is integrated with other system components and hence a set of system validation and integration tests is performed. System testing comes into picture after the Unit and Integration tests. It concentrates on testing the completely integrated functional system with a variety of techniques and methods. System testing is a series of different tests whose main objectives are to fully exercise the computer-based system. It allows the tester to prove that the system meets all objectives and system requirements. System testing is a verification process, which provides an external view of the system. This type of testing should be approached in a manner that would be easy for the user to use, since they are not concerned with how well the system works, responds, and the interfaces with its all the integral components. The user, in general, only wishes to know that the system functions properly. This type of testing is based on system requirements, an architecture document, or an operations document.

System testing is typically conducted by the testing team, under the supervision of the test team leader. Test leader is responsible for ensuring that adequate system testing is performed using appropriate testing techniques. The planning of system testing is the responsibility of the test team leader and should be developed with reference to the overall development and testing plan for the application under test. The following testing techniques are appropriate for system testing:

- *Black Box Testing* against high-level system requirement.

- *Thread Testing* against the high-level business requirements for the application under test.
- *Static Testing* including volume, stress, and performance testing.

System testing verifies the system when the testing occurs in a non-live test environment using dummy test data and could also be referred to as *verification testing*. Although each test has a different role, all work should verify that all system elements have been properly integrated and form allocated functions. The steps performed during software design and testing improve probability of successful software integration in the larger system. The following approach should be followed in conducting system testing:

- Review the requirements for the system and identify the high-level business requirements, the requirements for system performance, backup and recovery requirements, and security requirements.
- Review the environment in which the live system will run in order to identify any interoperability or compatibility issues with other systems.
- Examine the requirement for testing system procedures, system documentation, back-up and archive.
- Generate a set of test cases to test the system.

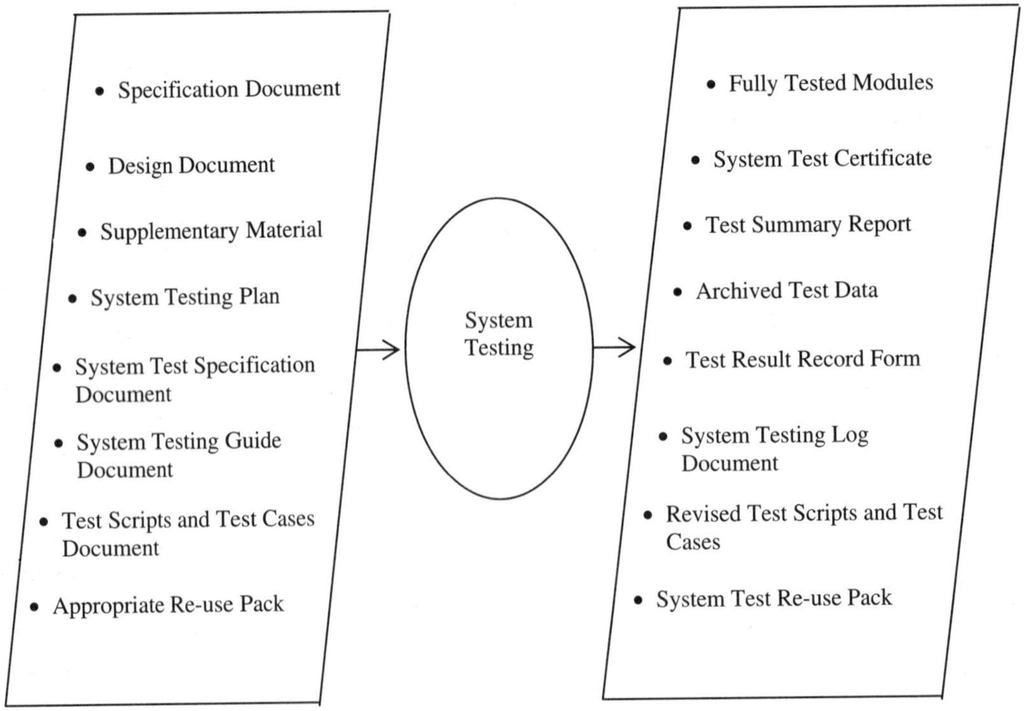


**Figure 6.4.4: System Testing: Inputs-outputs**

Figure 6.4.4 depicts the inputs to System testing before test execution begins, and the generated outputs from System testing. The general and typical popularly known System testing is described in the following section.

**Recovery Testing:** Generally, computer-based systems need to recover from faults and resume processing within a particular time. Recovery testing is a system test that focuses the software to fall in a variety of ways and verifies that recovery is properly performed. In certain cases, a system needs to be fault-tolerant. In other cases, a system failure must be corrected within a specified period of time or severe economic damage will happen. If it is automatic recovery then re-initialization, check pointing mechanisms, data recovery and restart should be evaluated for correctness. If recovery requires human intervention, the mean-time-to-repair (MTTR) is evaluated to determine whether it is within acceptable limits.

**Security Testing:** Security testing attempts to verify that protection mechanisms built into a system will, in fact, protect it from improper penetration. During security testing, the tester plays the role of the individual who wants to enter the system. The tester may try to get passwords through external clerical approaches. He may attack the system with customized software and purposely produce errors as well as try to find the key to system entry. The role of the designer is to make entry to the system more expensive than that which can be gained.

**Stress Testing:** Stress testing executes a system in the demands resources in abnormal quantity, frequently or volume. During stress testing, following tests may be performed:

- Special tests may be designed to generate ten interrupts per second, when one or two is the average rate.
- Input data rates may be increased by an order of magnitude to determine how input functions will respond.
- Test cases may be generated that require maximum memory or other resources may cause excessive hunting for disk-resident data, and thrashing in a virtual operating system.

### 6.4.5 Regression Testing

Regression testing checks the effect of changes made in the code. Most of the times, just before making a release of software to end user, software testing team is asked to check for the last minute changes in the code. In this situation, the only area of code, which is being affected by the change, needs to be checked. Regression testing helps in checking the only affected area of the code with the input from the development team. Regression testing is retesting unchanged segments of the application. It involves rerunning tests that have been previously executed to ensure that the same results can be achieved currently as were achieved when the segment was last tested. Regression testing is initiated after a programmer has attempted to fix a recognized problem or has added source code to a program that may have inadvertently introduced errors. One of the major objectives of the Regression testing is to ensure that the unchanged system segments function properly. It also ensures that the previously prepared manual procedures remain correct after the changes have been made to the application system. Regression testing is responsible for verifying whether the data dictionary of data elements that have been changed is correct.

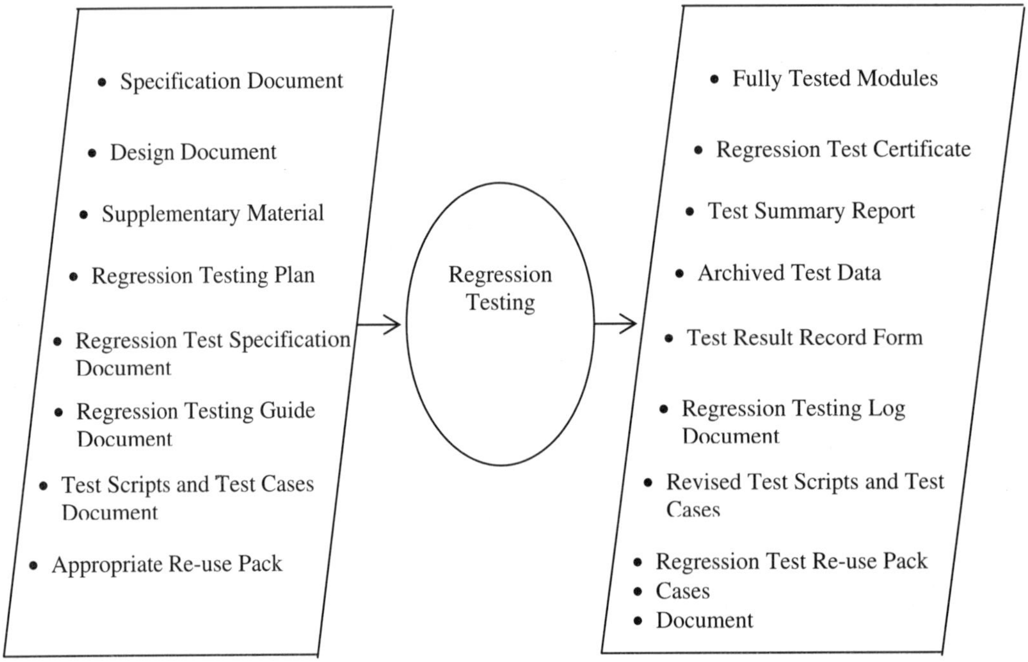


**Figure 6.4.5: Regression Testing: Inputs-outputs**

In fact the regression testing is the testing in which maximum automation can be done. But the extent of automation depends on whether the test cases will remain applicable over time. It is a quality control measure to ensure that the newly modified code still complies with its specified requirements and that unmodified code has not been affected by the maintenance activity. The process of regression testing is simple. The test cases that have been prepared already can be used. Expected results are also known. If the process is not automated it can be very time-consuming and tedious operation. It may be commonly performed in variety of ways including the following:

- Rerunning of previously conducted tests;
- Reviewing previously prepared manual procedures; and
- Comparing the current test results with the previously executed test results.

For Regression testing associated with the system, systems integration, or acceptance testing, the test team leader supervises the testing process and is responsible for ensuring that adequate Regression Testing is performed using appropriated testing techniques. Test analyst is responsible for obtaining and reviewing the re-use packs from previous testing phases for the purposes of selecting an appropriate set of test scripts and test cases to rerun against the new version of the software. It is assumed that Regression testing will take place either in a dedicated test environment or in the live environment. Figure 6.4.8 depicts the inputs to Regression testing before test execution begins, and the generated outputs from Regression testing.

## 6.5 COMPLETION OF TESTING

No doubt software testing is potentially endless process. All the defects need to be unearthed and removed to achieve complete testing, which is simply impossible. Practically, at some point, there is a need to stop testing and ship the software to the end user. But the major issue is at what point or when to stop the testing and ship the software. Realistically, testing is a trade-off between budget, time and quality. It is driven by profit models. The most common and pessimistic approach to stop the testing is to verify the allocated resources. Whenever some or any of the allocated resources time, budget, or test cases is exhausted, testing process is stopped. While, the optimistic approach is to stop testing when either reliability meets the requirement or the benefit from continuing testing cannot justify the testing cost. This approach generally requires the use of reliability models to evaluate and predict reliability of the software under test. Figure 6.5 shows the repeated running of evaluation.

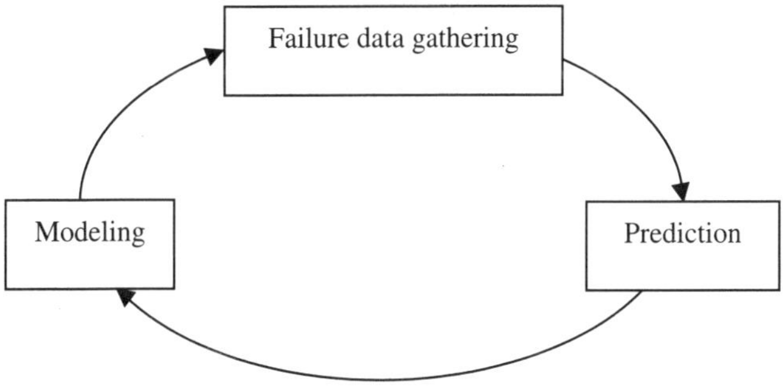


**Figure 6.5: Repeated Running of Evaluation**

## 6.6 SOFTWARE COMPONENT TESTING

Software reuse is being widely advocated and being accepted in software industry today. With the increase of software component products in commercial market, a new approach to develop large, complicated software application system based on available and reusable components has been introduced and known as *Component-Based Software Engineering*. Developed component based software system's quality depends on component quality, and any defective component may cause a ripple impact on overall system built on it. To validate component quality, a cost-effective test process needs to be performed.

Software component testing refers to testing activities that uncover software errors and faults to validate the quality of software components at the unit level. Unit testing of Module is the first step towards software testing process. It aims to check the quality of a module by performing unit test to uncover its program errors. Major objective of unit testing is to validate and confirm the quality of a software component by checking its functions, behavior and performance. This will assure that the module under test meets its functional requirements and specified design in a given operational environment. Software component may be tested by component venders as well as by the users in *Component-Based Software Engineering*. Two typical ways of component testing are described briefly as follows:

**Vendor-Oriented Component Testing:** In a vendor-oriented component testing, testing activities are performed by the component venders to validate a software component based on its specifications. Vendor-oriented component testing checks whether the developed components are of high quality based on the given specifications or not. Another objective is to verify whether the developed component is based on the specified standards and component models or not. Component developers will be responsible to answer these questions.

**User-Oriented Component Testing:** User-oriented component testing involves a component validation process to make sure that every involved software component delivers the specified functions, interfaces, and performance. User-oriented component testing will check whether the selected component is reusable component and is right for a system or not. Component users will be responsible to answer for the correct use of reusable component in a system and correct update of component for the project.

## 6.7 TESTING REAL TIME SYSTEMS

Testing is an important part of any software development project, and can typically surpass more than half of the development cost. Testing is a fundamental step in any development process. It consists of applying a set of experiments to a prototype system, with multiple aims, from checking correct functionality to measuring performance. For testing of sequential software it is usually sufficient to provide the same input in order to reproduce the output. For real-time systems, on the other hand, it is required to control, or observe, the timing and order of the inputs. If the system is multitasking, it is also needed to take timing and the concurrency of the executing tasks into account. Real-time systems have to respond to external interactions in a predetermined very small amount of time. Successful completion of an operation depends upon the correct and timely operation of the system.

Reproducible and deterministic testing of sequential programs can be achieved by controlling the sequence of inputs and the starting conditions, that is, for the given same initial state and inputs, the sequential program should deterministically produce the same output on repeated executions, even in the presence of systematic faults. Whenever regression testing is performed, reproducibility becomes essential. In reproducibility, the same test cases are run repeatedly with the intent to validate that either an error correction had the desired effect, or simply to make it possible to find the error when a failure has been observed. However, trying to directly apply test techniques for sequential programs on distributed real-time systems is bound to lead to non-determinism and non-reproducibility, because control is only forced on the inputs, disregarding the significance of order and timing of the execution and communication tasks. Real-time systems including embedded systems, communication protocols, multimedia systems, operate in an environment with strict timing constraints. Testing is even more important for safety-critical computer based systems due to stringent reliability and safety requirements. Real-time program testing involves additional problems caused by the time-dependent interaction of operational programs, control program, complex hardware, and random input to the system.

However, most safety-critical computer based systems are real-time systems. Adequate pre-installation testing is important for testing real-time system because errors are difficult to isolate and correct once such systems are operational. While testing real-time systems, two important facts should be considered. First, it is not sufficient to check for the correct output, the system under test should also be checked for the time of the output to be correct. Second, which output will be produced and what is the timing of the produced output, will be determined by the timing of inputs.

Several formal models have been proposed in literature for real-time system such as finite state machine, Petri Net etc. Verification and testing are two different techniques that are usually used to cope with the correctness of a system. Verification deals with the specification of the system under consideration and aims to ensure that the designed specification satisfies predefined functional and timing requirements. Testing is an important activity, which aims to ensure the quality of the implementation. Testing procedure consists of generating test suites and applying them to the implementation. The effort of developing testing routines is commensurate with that required for the control program, and the volume of code in the routines can be expected to exceed that of the control program.

## 6.8 MODELS FOR SOFTWARE TESTING

In software testing, a model may be a representation of an idea, a diagram (like an Entity-Relationship diagram), or a physical representation. A model helps to think of a system or concept in a certain ways and as a representation is not going to be perfect or complete. Modeling applications helps to imagine areas of potential problems based on experience and technical knowledge of the various technologies used. It also helps to narrow down or get repeatable cases for bugs found in the application, and to generate different testing ideas instead of one happy path based on requirements documents. There are various models of software testing, including V-Model, W-Model, and B-Model described in the following section.

### 6.8.1 V-Model

The V-Model was originally developed from the waterfall software process model. V describes the pictorial arrangement of the individual phases. The V is also a synonym for verification and validation. V-Model defines tangible phases of the software development process. It proposes a logical sequence in which the identified phases should be approached, and defines logical relationships between the phases. V-Model demands the written testing documentation as soon as possible, and gives equal weight to development and testing. V-Model provides a simple and easy to follow map of the software development process

There are four process phases in V model:

- Requirement Analysis
- Specification
- High Level Design
- Detailed Design

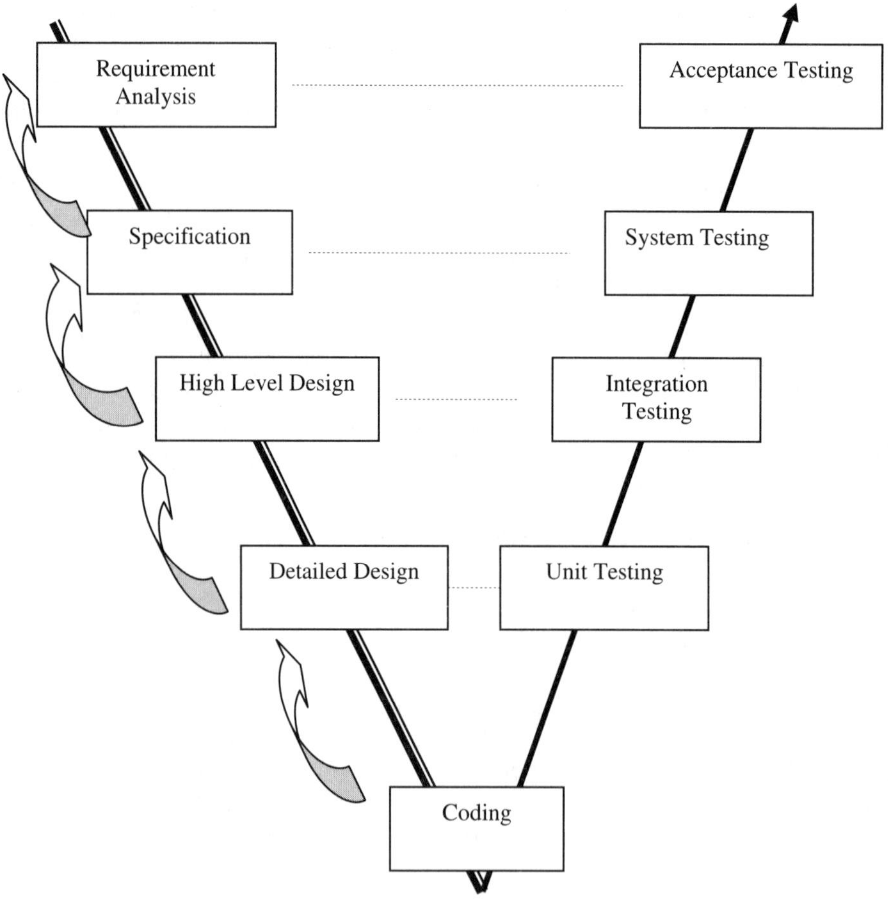


**Figure 6.8.1: V-Model**

All these process phases have a corresponding verification and validation based testing phase. Detailed design of modules is tested by unit testing. Integration testing validates high level design of system, while specifications are tested by system testing. Finally acceptance testing verifies the requirements. V-Model gets its name from the timing of the phases. Starting from the requirements, the system is developed one phase at a time until the lowest phase is finished. Now testing begins, starting from unit testing and moving up one test level at a time until the acceptance-testing phase is completed as shown in figure 6.8.1. The figure is self-explanatory and easy to understand. However, this model has been criticized for not taking ease of the complexities involved, and for misleading project managers if relied upon.

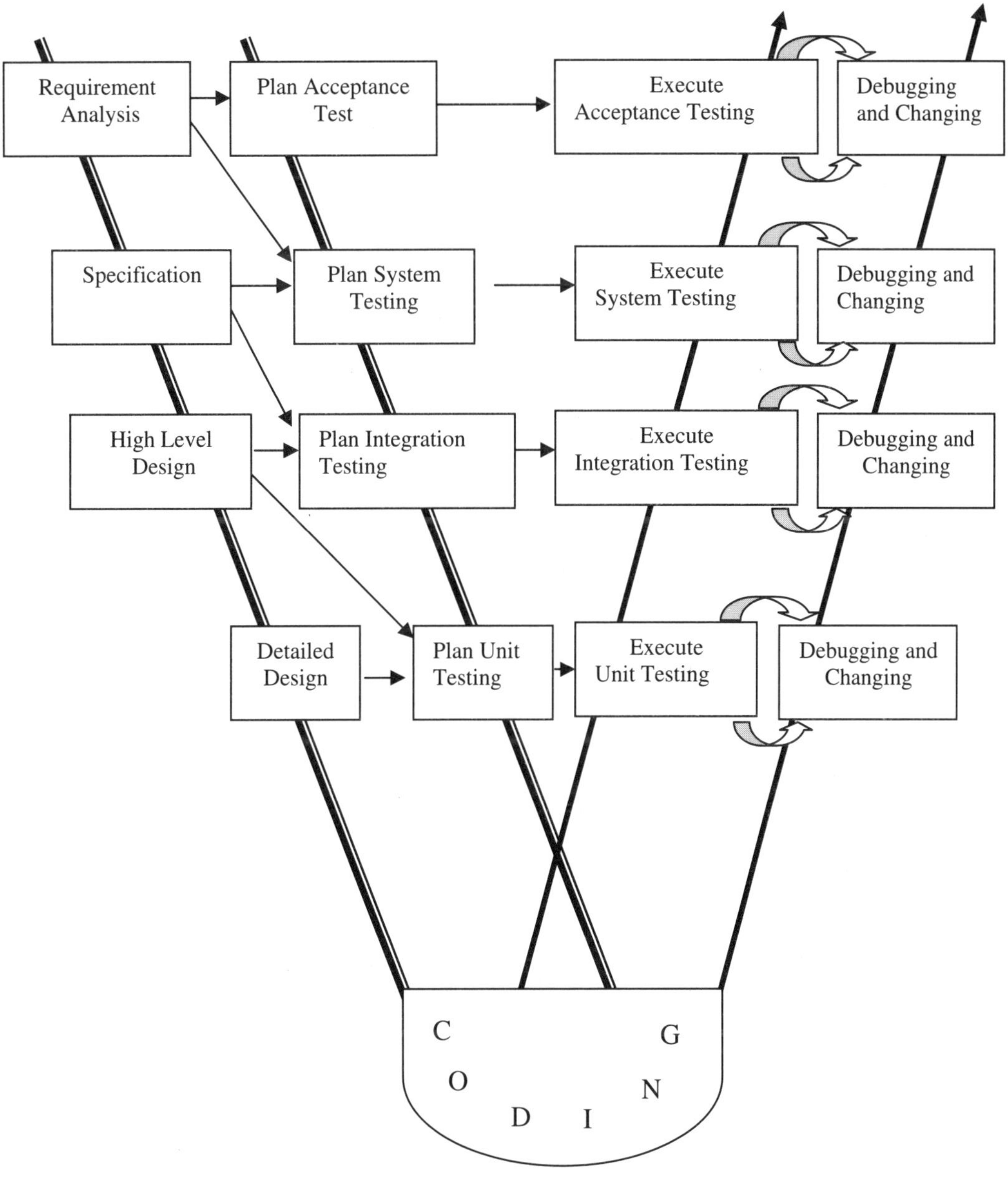


**Figure 6.8.2: W-Model**

### 6.8.2 W-Model

Generally models adopted by the developers termed to be deficient in various ways. In most of the cases, test activities start after the implementation. Generally connection between the various test stages and the basis for the test is not clear. Above all none of the models available take care of the tight link between testing, debugging and change tasks during the testing phase.

All these disadvantages led to development of another model called W-model. W-Model is based on the general V-model. Generally test process receives too little attention in the models presented. In order to place testing on an equitable footing, a second 'V' dedicated to testing is integrated into the model, which forms the basis of W-Model. W-Model is shown in figure 6.8.2.

The importance of the tests and the ordering of the individual activities for testing are clear in W-Model. In the W-model it is clear that the division between constructive tasks on the left-hand side and the more supplementary tasks on the right-hand side. V-mode does not strive for closer co-operation between development and testing activities that must exist for effective and efficient software development practices. The model clearly emphasizes the fact that testing is much more than just construction, execution and evaluation of test cases. W-model appears closer to real practice when the test expenditure is given 40 percent and more, and may be desirable for implementation when software complexities are rapidly increasing. However, it appears that the different activities have an equal requirement for resources such as time, personnel, etc. which may not be commonly true in practice.

### 6.8.3 B-Model

Another important model for testing software products is B-Model that is, Butterfly Model. Three portions of this model give the structure of Butterfly. The first portion of the model relates to the left wing of butterfly and depicts the Test Analysis. While the second portion relates to the right wing and represents Test Design. Finally, the third portion, the body of the butterfly is nothing but the Test Execution. Figure 6.8.3 shows the generic Butterfly Model.

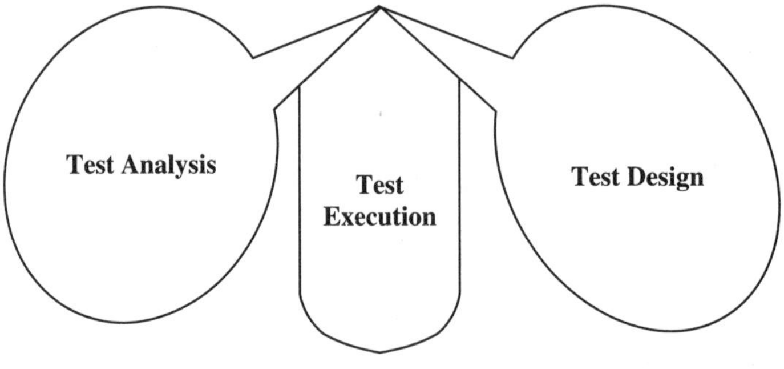


**Figure 6.8.3: B-Model**

Analysis is the key factor, which drives effective planning. During Test Analysis the required documents are carefully studied by the Test Personnel, and the final analysis report is documented including the software requirement specification, functional specification, architecture document and use-case dosuments. The analysis report would consist of the understanding of the application, the functional flow of the application, number of modules involved and the effective test time. The analysts will verify that each requirement is tagged in a manner that allows correlation of the tests for that requirement to the requirement itself. During test analysis, the analysst creates a test-sketch that includes the tentative approach and indicates the test objectives.

Test design is another important component of the butterfly model. It represents the act of designing and implementation of the test cases needed to verify the design artifact as replicated in the implementation. Test design is an exercise in the prevention, detection, and elimination of bugs in software. The outputs of the test analysis phase are the foundation for test design. Its focus is not to assimilate information created by others, but to implement procedures, techniques and data sets that achieve the test objectives. During test execution, the test cases designed are put to operation. These test cases need to be accurate, economically executable, repeatable, appropriate to the task, and traceable enough. After executing the test cases, test execution report and defect report are delivered as a final deliverable.

## 6.9 CONCLUSION

A strategy for software testing integrates software test-case design techniques into a well-planned implementation-structure that causes the production of software. It provides a road-map for the software developer, the quality assurance organization, and the customer. Generic testing strategy includes test planning, test case design, test execution, and the resultant data collection evaluation. No doubt, a test strategy is a vital enabler to the process - keeping focus on core values and consistent decision-making to help achieve desired goals with best use of resource. Independent testers are responsible to remove the conflicts of interest inherent when the builder is testing his or her own product. One of the most important goals of software testing is verification and validation. Software verification and validation is one of the broader domains for software testing. Testing can serve as qualitative metrics, and is heavily used as a tool in the verification and validation process. A strategy for software testing may be to move upward along the spiral, with unit testing at the vortex of the spiral. Such strategy calls for concentrating on each unit of the software as implemented by the source code. The unit test is the lowest level of testing performed during software development. Unit test case documents may be checked by a unit test case checklist.

Quality assurance analysts and peer programmers are responsible for reviewing the program specifications and unit test cases. Integration testing is a systematic approach that produces the program structure, and test to identify errors associated with interfacing, and deals with many communicating modules. Integration based faults are ultimately caused by incomplete specifications or faulty implementations of interfaces, resource usage, or required properties. Top-Down integration testing is an incremental approach to construction of program structure, while bottom-up integration testing begins with testing the modules at the lowest level which is said to be atomic modules. The major disadvantage of Top-Down approaches is the need for stubs and the difficulties that are linked with them. On the other hand, the major drawback of Bottom-Up integration is that the program does not exist until the last module is included. Validation testing is started as a final set of software tests, and is achieved through a series of Black-Box tests that show conformity with requirements. One of the important elements of the validation process is a configuration review. Configuration review ensures that all the components of the software configuration have been properly developed. Alpha testing is performed at the developer's sites, with the developer checking over the customers shoulder as they use the system to determine errors, while Beta testing is a detailed testing and needs to cover all the functionalities of the product and also the dependent functionality testing.

System testing also known as verification process that provides an external view of the system. It should be approached in a manner that would be easy for the user to use. Recovery testing focuses the software to fall in a variety of ways and verifies that recovery is properly performed. Security testing attempts to verify that protection mechanisms built into a system will protect it from improper penetration, while regression testing is the testing in which maximum automation can be done. Verification of the allocated resources is the most common and pessimistic approach to stop the testing, and testing process used to be stopped whenever some or any of the allocated resources are exhausted. Software component testing uncovers software errors and faults to validate the quality of software components at the unit level. V-Model proposes a logical sequence in which the identified phases should be approached.

## KEY TERMS

Strategy, Testing Strategy, Verification, Validation, Unit Testing, Integration Testing, System Testing, Completion of Testing, Component Testing, Functional Technique, Structural Technique, Error Based Technique, Input Domain Testing, Equivalence Partitioning, Boundary Value, Syntax Checking, Special Values, Output Domain Testing, Statement Testing, Branch Testing, Conditional Testing, Expression Testing, Path Testing, Fault Seeding, Mutation Testing, Historical Test Data, Top-Down Approach, Bottom Up Approach, Alpha Testing, Beta Testing, Security Testing

## REVIEW QUESTIONS

## OBJECTIVE TYPE QUESTIONS

1. Verification can be done with________.
   (a) Walkthrough only
   (b) Inspection only

(c) Both Walkthrough and Inspection
(d) None of the above

2. Dependability property can be verified by using________.
(a) Functional testing technique
(b) Model-based testing technique
(c) Both functional and model based testing technique
(d) None of the above

3. Unit testing concentrates ________on the smallest elements of program.
(a) Verification
(b) Validation
(c) Verification and Validation
(d) None of the above

4. Information, such as execution traces timing profile and test coverage, is provided by________.
(a) Dynamic analysis
(b) Static analysis
(c) Code review
(d) None of the above

5. ________investigates the textual source of software, looking for problems and gathering metrics.
(a) Dynamic analysis
(b) Static analysis
(c) Code review
(d) None of the above

6. Program specification and unit test case are reviewed by________.
(a) Quality Assurance Analysts
(b) Peer programmer
(c) Both quality assurance analysts and peer programmer
(d) None of the above

7. Integration testing strategy can be classified as________.
(a) Structural
(b) Feature-driven
(c) Structural and feature driven both
(d) None of the above

8. Structural strategies include________.
   (a) Top-down approach
   (b) Bottom-up approach
   (c) Both top-down and bottom up approach
   (d) None of the above
9. Software technical contents are evaluated normally through________.
   (a) Alpha testing
   (b) Beta testing
   (c) Both alpha and beta testing
   (d) None of the above

## SHORT ANSWER TYPE QUESTIONS

1. What do you mean by testing strategy? What are the characteristics of good testing strategy?
2. Write the need, importance and significance of software testing strategy.
3. Write down the integral constituents of test strategy.
4. 'A software testing strategy provides a road map for software developers'. Comment categorically.
5. 'Diverse testing techniques are appropriate at diverse points in time'. Justify reasonably.
6. What is the purpose of considering verification issues during testing?
7. What is the purpose of considering validation issues during testing?
8. Distinguish between dependability properties and usability properties, clearly with appropriate examples.
9. What do you mean by control path? List the advantages of testing control path.
10. 'Unit tests have to be repeated whenever software is modified'. Comment.
11. List the basic activities associated with unit testing?
12. How static analysis does investigate the textual source of software? Explain in brief.
13. How dynamic analysis does provide information such as timing profile and test coverage? Explain in brief.
14. Distinguish between program specification and technical specification.
15. How programmer does prepare program specification for unit? Describe the basic steps in brief.
16. Distinguish between fault seeding and mutation testing.

17. How are mutants created? Explain with an appropriate example.
18. What do you mean by integration faults?
19. What are the integration testing strategies? List and explain in brief.
20. List the objectives of integration testing, and its significance.
21. Prepare a probable list of the non-functional problems that may cause integration faults.
22. How does consistent interpretation of parameters cause integration fault?
23. Distinguish between the top-down and bottom-up approach with respect to integration testing.
24. Writing stubs for untested models are very expensive. Comment.
25. Distinguish between concepts of are negative and fare positive.
26. What are the advantages and disadvantages of bottom-up approach of integration?
27. How does configuration review assist in validation process?
28. How does software component testing deviates from testing traditional software?
29. Differentiate between vendor-oriented test and user-oriented component test.

## DESCRIPTIVE QUESTIONS

1. How does software test strategy help in outlining what to plan, and how to plan? Discuss with appropriate case example.
2. 'A software testing strategy should be flexible enough'. Justify.
3. Do an exhaustive review on verification and validation. Critically examine the issues related to verification and validation.
4. Write the goals of the unit testing, and suggest guideline to achieve these goals.
5. How does programmer prepare unit test case document for the particular unit? Describe the methodology with an example.
6. Propose a comprehensive checklist to be used to check the completeness of unit test case documentation.
7. Integration faults are ultimately caused by incomplete specifications. Justify.
8. Propose a checklist for verifying the completeness of test process.
9. Prepare a comprehensive analysis report on different models for testing.
10. Do a critical review on B-Model of software testing. Find out the limitations and delimitation of this model.
11. Write a paper on software testing strategic issues.

## REFERENCES

1. Roger Ferguson & Bogdan Korel, "Software Test Data Generation using the Chaining Approach", International Test Conference, IEEE.
2. S.P. Ng, T. Murnane, K. Reed, D. Grant, T.Y. Chen, "A Preliminary Survey on Software Testing Practices in Australia", Proceedings of the 2004 Australian Software Engineering Conference (ASWEC'04) 2004 IEEE
3. Timothy J. Shimeall, Timothy J. Kelly, "Examination of a Software-Test Cost Model", 1994 PROCEEDINGS *Annual* RELIABILITY and MAINTAINABILITY Symposium, pp:26-30.
4. JoHo W. Cangussu, Modeling and Controlling the Software Test Process, 2001 IEEE
5. T. Y. Chen & P L Poon, "Teaching Black Box Testing",
6. Jotio Cangussu, Raymond DeCarlo, and Aditya Mathur. A state variable model for the software test process. In Proceedings of 13th International Conference on Software & Systems Engineering and their Applications( ICSSEA), Paris-France, December 2000.
7. G. Cugola and C. Ghezzi. Software processes: A retrospective and a path to the future. Sofiware Process Improvement and Practice, 1999.
8. Lennart Ljung. System idenrijkation: theory for the user. Prentice-Hall, Englewood Cliffs, NJ, 1987.
9. David G. Luenberger. Introduction to Dynamic Systems: theoty, models and applications. John Wiley & Sons, 1979.
10. IEEE Glossary of Sofhvare Engineering Term, The Institute of Electrical and Electronics Engineers Press, 1991.
11. IEEE Standard: Soware Tesr Documentation, The Institute of Electrical and Electronic Engineers Pres, 1983.
12. Booch. G., Sofhvare Components with Ado, Benjamin/Cummings Publishing, 1987.
13. Camuffo, M., MaiocAi, M., Morselli, M., "Automatic Software Test Generation".
14. Kelly, T. 1.. "An Evaluation of a Test Scheduling Solution", M.S. Thesis, Computer Science Dept., Naval Postgraduate School, Monterey, CA, March 1993.
15. Myers, G. J., The Art ofSo@are Testing, John Wiley & Sons, 1979.
16. Shimeall, T. J. and Leveson, N. G., "An Empirical Comparison of Software Fault Tolerance and Fault Elimination", IEEE Transactions on Softwure Engineering, February 1991.
17. P. Ammann, "System Testing via Mutation Analysis of Model Checking Specifications", ACM SIGSOFT Software Engineering Notes, Volume 25, No. 1, January 2000, pp 30.
18. Murial Daran, Pascale Thévenod-Fosse, "Software Error Analysis: a Real Case Study involving Real Faults and Mutations", ACM SIGSOFT Software Engineering Notes, Volume 21, No. 3, May 1996, pp 168-171.

19. L. Groves, R. Nickson, G. Reeves, S. Reeves and M. Utting "A Survey of Software Practices in the New Zealand Software Industry", Proceedings of the.2000 Australian Software Engineering Conference, Gold Coast, Queensland, Australia, April 28-30, 2000 pp 189-101.

20. B. Kitchenham and S. L. Pfleeger, "Principles of Survey Research, Part 5: Populations and Samples", ACM SIGSOFT, Software Engineering Notes, Volume 27, No. 5, September 2000, pp 17-20.

21. J. D. Musa and A. F. Ackerman, "Quantifying Software Validation: When to Stop Testing?", IEEE Software, May 1989, pp 19–27.

22. N. Mateev, V. Menon, K. Pingali, "Fractal Symbolic Analysis", Proceedings of the 15th international conference on Supercomputing, June 2001, pp 38-49.

23. T. Murnane and K. Reed. "On the Effectiveness of Mutation Analysis as a Black-box Testing Technique", Proceedings of the 2001 Australian Software Engineering Conference, Canberra, Australia, 2001, pp 12-20.

24. S. L. Pfleeger, *Software Engineering: Theory and Practice*, Prentice-Hall, USA, 2001.

25. R. S. Pressman, Software Engineering: A Practitioners Approach, McGraw-Hill, International Edition, 1992.

26. Bhandari, M. J. Halliday, J. Chaar, K. Jones, J. Atkinson, C. Lepori-Costello, P. Y. Jasper, E. D. Tarver, C. C. Lewis, and M. Yonezawa. In-process improvement through defect data interpretation. The IBM System Journal, 33(1):182–214, 1994.

27. B. W. Boehm. Software Engineering Economics. Prentice Hall, Englewood Cliffs, NJ, 1981.

28. J.K. Chaar, M.J. Halliday, I.S. Bhandari, and R. Chillarege. In-process evaluation for software inspection and test. IEEE Transactions on Software Engineering, 19(11):1055–1070, November 1993.

29. D. M. Cohen, S. R. Dalal, M. L. Fredman, and G. C. Patton. The AETG system: An approach to testing based on combinatiorial design. IEEE Transactions on Software Engineering, 23(7):437–444, July 1997.

30. P. G. Frankl and O. Iakounenko. Further empirical studies of test effectiveness. In Proceedings of the ACM SIGSOFT 6th International Symposium on the Foundations of Software Engineering (FSE-98), volume 23, 6 of Software Engineering Notes, pages 153–162, New York, Nov. 3–5 1998. ACM Press.

31. T. Graves, M. J. Harrold, J.-M. Kim, A. Porter, and G. Rothermel. An empirical study of regression test selection techniques. In Proceedings of the 20th International Conference on Software Engineering, pages 188–197. IEEE Computer Society Press, April 1998.

32. P. A. Hausler, R. C. Linger, and C. J. Trammell. Adopting cleanroom software engineering with a phased approach. IBM Systems Journal, March 1994.

33. Independent Assessment Team. Mars program independent assessment team summary report. Technical report, 2000.

34. K. Reed, "Testing, testing, testing", Australian Computer Society Information Age, April/May 2002, pp 56–58.

35. D. S. Alberts, "The economics of software Proceedings: 1976 National Computer Conference, vol. 45, Montvale, N.J., pp. 433-442

36. R. DeMillo, W. M. McCracken, R. J. Martin,

37. J. F. Passafiume, Software Testing and Evaluation, The Benjamin/Cummings Publishing Company, 1987.

38. D. C. Ince, "The automatic generation of test data," The Computer Journal, Vol. 30, No. 1, P I 1987, pp. 63-69.

39. L. A. Clarke, "A system to generate test data and symbolically execute programs," IEEE

40. Trans. on Software Eng., vol. SE-2, No. 3, Sept. 1976, pp. 215-222.

41. B. Korel, "Automated test data generation," IEEE Transactions on Software Engineering, vol. SE-16, NO. 8, August 1990, pp. 870- 879.

42. W. E. Howden, "Symbolic testing and the DISSECT symbolilc evaluation system," IEEE Transaction on Software Engineering, vol. 4, No. 4, 1977, j~p. 266-278.

43. Jaaksi. Assessing software projects: Tools for business owners. In Proceedings of the 9th European Software Engineering Conference held jointly with 10th ACM SIGSOFT International Symposium on Foundations of Software Engineering (ESEC/FSE 2003), pages 15–18. ACM Press, September 2003.

44. J.-M. Kim and A. Porter. A history-based test prioritization technique for regression testing in resource constrained environments. In Proceedings of the International Conference on Software Engineering (ICSE 2002), May 2002.

45. J. L. Lions. Ariane 5, flight 501 failure, report by the inquiry board. Technical report, 1996.

46. B. Marick. The Craft of Software Testing: Subsystems Testing Including Object-Based and Object-Oriented Testing. Prentice-Hall, 1997.

47. J. Nielsen. Designing Web Usability: The Practice of Simplicity. New Riders Publishing, Indianapolis, 2000.

48. T. J. Ostrand and M. J. Balcer. The category-partition method for specifying and generating functional tests. Commun. ACM, 31(6):676–686, June 1988.

49. B. Korel, "A Dynamic Approach of Automated Test Data Generation," Conference on Soft ware Maintenance, San Diego, CA, November 26-29, 1990, pp. 311-317.

50. P. Gill, Numerical Methods for Constrained Optimization, New 'fork: Academic, 1974.

51. H. Glass, L. Coopea, "Sequential search: a method for solving constrained optimization problems," Journal of ACM, vol. 12, No. 1, Jan. 1965, pp. 71-82.

52. Luciano Baresi, Mauro Pezz, "An Introduction to Software Testing", Electronic Notes in Theoretical Computer Science 148 (2006) 89–111

## USEFUL LINKS

http://www.sei.cmu.edu/publications

http://www.Stickyminds.com

http://www.sasystems.com

http://www.softwareqatest.com

http://www.eng.mu.edu/corlissg/198.2001/KFN_ch11-tools.html

http://www.ics.uci.edu/~jrobbins/ics125w04/nonav/howto-reviews.html

http://www.processimpact.com

http://www.webopedia.com

# Chapter 7

# Planning for Software Testing

*The Plan is nothing, the planning is everything.*
**Dwight D. Eisenhower**

## OBJECTIVES

A successful completion of this chapter will enable the reader for the following:

- To realize the benefits of software test planning.
- To understand role of software test plan.
- To prepare a test plan specification.
- To integrate software testing with development life cycle.
- To understand master test plan to conduct test.
- To formulate detailed phase wise test plan in advance.
- To setup the schedule to conduct and windup phase wise testing.
- To understand levels of test planning.
- To identify the necessary acquisitions for testing.
- To realize the need and importance of acquisitions in preparing test plan.
- To elaborate test items to be identified in advance to conduct test.
- To understand the role of requirement specifications in preparing test plan.
- To prepare design specifications in advance to conduct test.
- To identify set of test deliverables as a result.
- To finalize phase wise testing strategy.
- To identify the technical specifications in advance to conduct test.
- To realize the need and importance of documentation in test specification.
- To set the criteria for successful testing.
- To identify what to test or not to test.
- To develop test plan for each and every phase.

## 7.1 INTRODUCTION

Testing is a process of analyzing a software items to detect the differences between existing and required conditions. Software test plan is designed to prescribe the scope, approach, resources and schedule of all testing activities. The plan must identify the items to be tested, the features to be tested, the types of testing to be performed, the personnel responsible for testing, the resources and schedule required to complete testing, and the risks associated with the plan. Testing will be performed at several points in the life cycle as the product is being built. Testing is very dependent activity and not as independent as normally it is considered to be. As a result, test planning is a continuing activity performed throughout the system development life cycle. Test plans should be developed for each phase of software development life cycle.

The most basic role of a test plan is to guide all testing activities. Efficient testing requires a substantial investment in planning, and a willingness to revise the plan dynamically to account for changes in requirements, designs, or code as bugs are uncovered. Software test plan defines what is to be tested and what is to be overlooked, how the testing is to be performed and by whom. It is therefore largely a managerial document. Therefore, the target audience of the plan should be a manager with a decent grasp of the technical issues involved. Test plans are made according to the details of the software implementation, such as programming language, logic, and styles. Ideally, test planning should take into account static as well as dynamic testing.

ANSI/IEEE Standard for SW Test Documentation defines Test Plan as "a document describing the scope, approach, resources, and schedule of intended testing activities". It identifies test items, the features to be tested, the testing tasks, who will do each task and any risks requiring contingency planning. A software test plan enables to map the tests to the requirements and define the entry and exit criteria for each phase of testing. It is to define the test-strategy for the system needed for testing, and the work product for each phase of testing. Software test plan enables to assess the test-related risk and a plan for their mitigation much before uncalled damages are done.

The fundamental planning process may be based on three phases including necessary acquisition, testability assessment, and phase-wise testability assessment. Necessary acquisition phase describes what is to be tested, and the pre-requisites to conduct the tests. Testability assessment verifies the level of testability of software under test on the basis of the availability of internal, external, and environmental facilitators. The third and last component of test planning is phase wise testability assessment of software under test, which includes strategic specifications, technical specifications, and operational specifications as subcomponents. Strategic specification deals with the selection of appropriate and suitable level specific approach to conduct the test. Technical specification verifies the availability of level specific test tools, and operational specification executes all of the test cases, and observes the results. That is, a model test plan should include precise specifications in all three dimensions in the very-context of the environmental specifications.

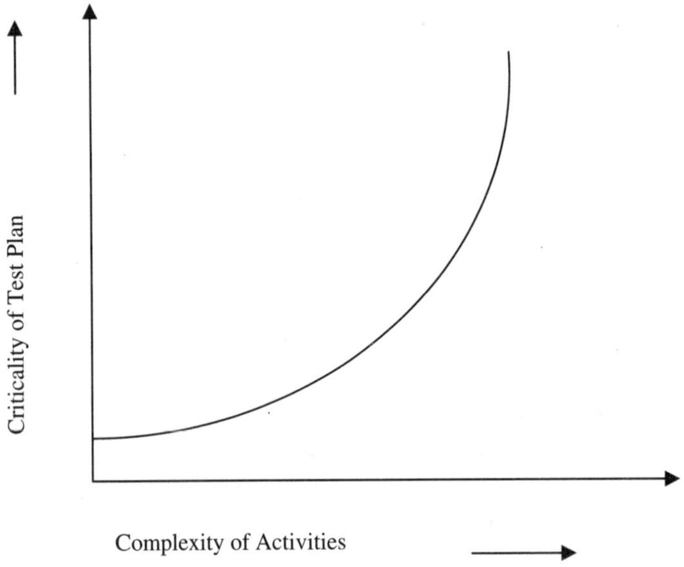


**Figure 7.1: Importance of Test Planning**

Deciding the number and scope of test plans required should be one of the first strategy decisions made in test planning. In test planning, even though the document is important. The process is ultimately more important than the document. Discussing issues of what and how to test early in the development life cycle can save a lot of time, effort, cost, and disagreement later. In spite of creating long list of test cases, the major objective of test planning includes the important issues of testing strategy, resource utilization, responsibilities, risks, and priorities. Test planning is the mental preparation, which improves the understanding of situation. As the complexity of a testing activity increases, the criticality of having a good Master Test Plan increases exponentially, as illustrated in figure 7.1.

## 7.2 TEST PLAN SPECIFICATION

To be most effective, test planning must start at the beginning and proceed in parallel with software development. General project information is used to develop the master test plan to assess testability of software as a whole, while more specific software information is used to develop the detailed phase wise test plan to predict phase wise testability of software. A *level* of test is defined by a particular environment, which is a collection of people, hardware, software, interfaces, data, and even the viewpoints of the testers. Creating a test plan for a specific *level* requires a clear understanding of the unique considerations associated with that level. Product risk issues, resource constraints, staffing and training requirements, schedules, testing strategy, and other factors must all be considered. Figure 7.2(a) depicts a test plan specification with the following components.

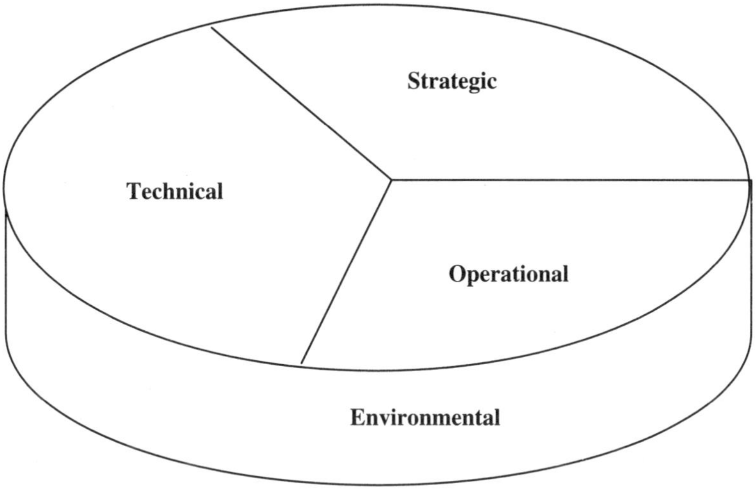


**Figure 7.2(a): Test Plan Specification**

**Environmental Specifications:** This section describes the software test environment at each intended test site as illustrated in figure 7.2(b). Documenting the physical components required for test execution helps to identify potential gaps in what is required and what actually exists. It is required to specify both the necessary and desired properties of the test environment, including the physical characteristics of the facilities like hardware, communications and system software, the mode of usage and any other software or supplies needed to support the test.

Test Environment is a stable area for independent systems and integration testing by the test team. This area consists of objects as they are completed by developers and meet the requirements for promotion. This environment ensures that objects are tested with the latest stable version of other objects that may also be under development. Software test environment also specifies the level of security that must be provided for the test facilities, system software, and proprietary components such as software, data, and hardware. It will identify one or more of following:

- Special test tools needed;
- Any other testing needs;
- Source for all needs that is not currently available to the test group.

The test environment is initially populated with the latest operational application and then updated with new changed objects from the development environment. An attempt should be made to configure the testing environment as similar to the real –world system as possible. If the system is designed to be run on multiple configurations, a decision must be made whether to replicate all of these configurations, only the riskiest, only the most common, or some other combination. In addition to specifying the hardware and software, it is also necessary to identify where the data will come form to populate the test database.

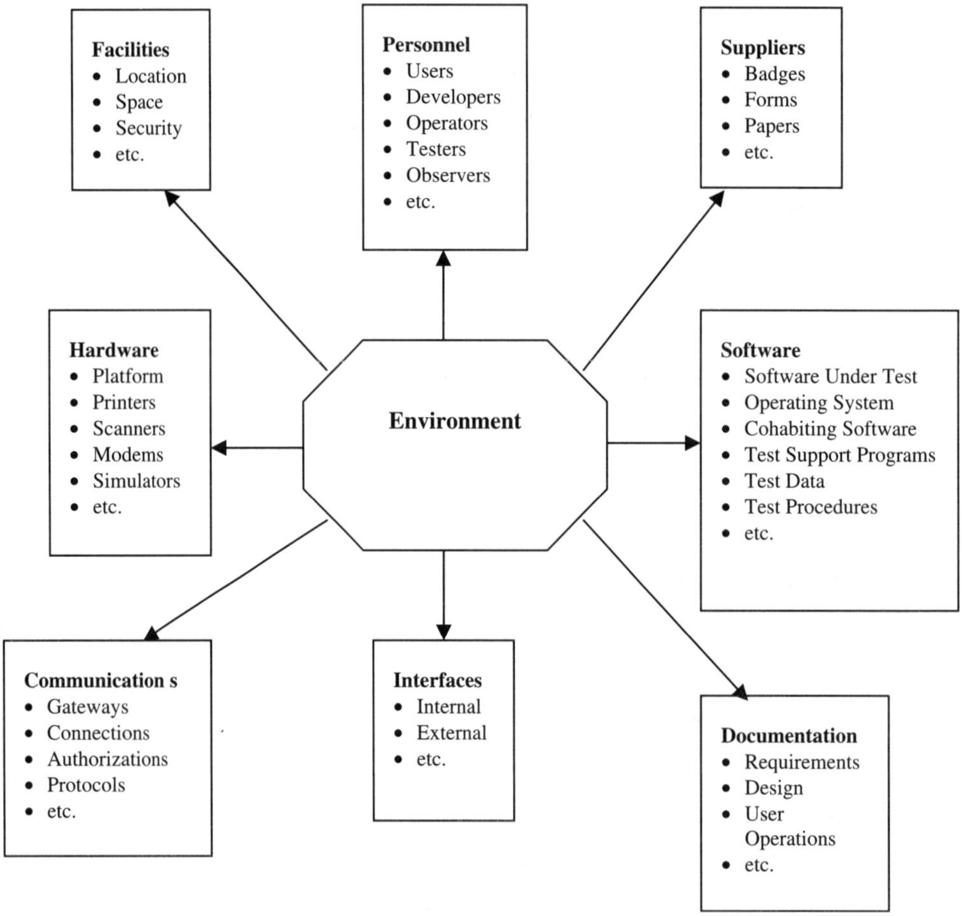


**Figure 7.2(b): Environmental Specifications**

**Strategic Specifications:** This section is the heart of the test plan, and should contain a description of how testing will be performed and explain any issues that have a major impact on the success of testing and ultimately on the project. Strategy to conduct test at each phase should be explained with the context of particular phase, including the entrance and exit criteria from one phase to another. Testing strategy describes the overall approach and techniques to be used in each phase during testing. It outlines the steps in the overall testing process. Strategic specification describes the phase wise testing approach to be followed.

It is necessary to identify all software features, combinations of software features to be tested and also the design specification associated with this. All the features and significant combinations of features that will not be tested from both the users' viewpoint of what the system does. Moreover, a configuration management/version control viewpoint will also be identified and the proper reasons justifying why not to test will be presented.

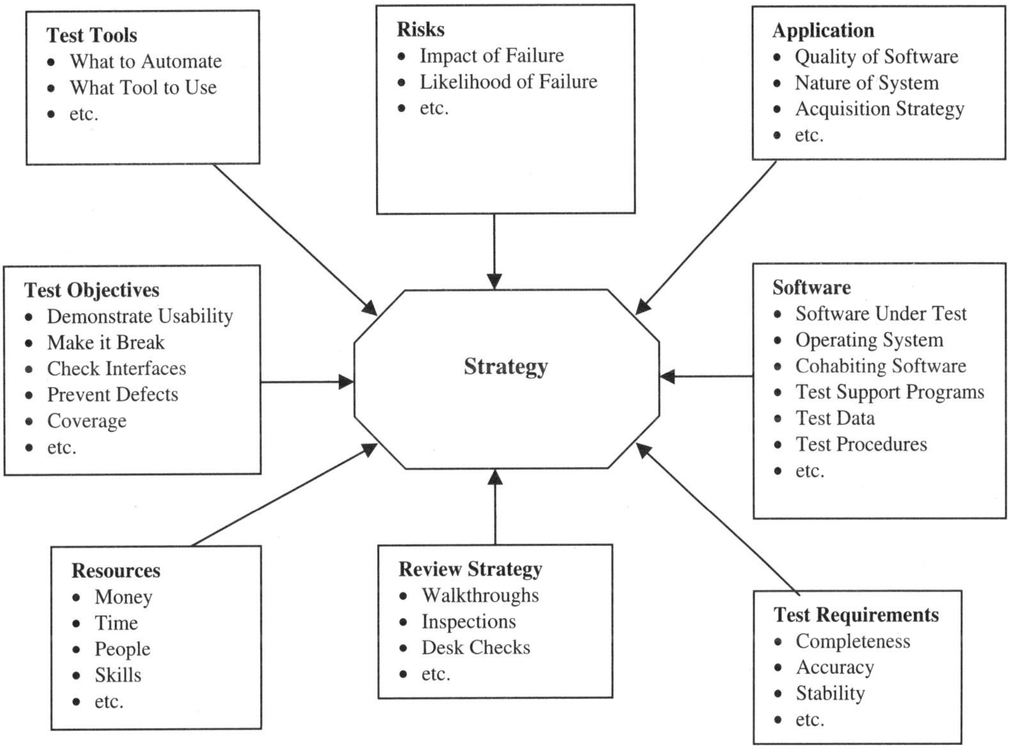


**Figure 7.2(c): Strategic Specifications**

The completion criteria for the overall test plan will be set in advance in strategic specification. It is required to set some benchmark or prepare a checklist at each phase to know whether the testing phase is complete or not. For example, the completion criteria for the integration test level may include checking whether all test cases are completed, a specified percentage of cases are completed with a percentage containing some number of minor defects, and code coverage tool indicates that all code is covered. Figure 7.2(c) illustrates some typical influences on strategic decisions.

**Technical Specifications:** Another major issue in test plan specification that should be addressed is the selection of appropriate tools and automation. Testing tools can be a tremendous help to the development and testing staff. But they can also spell disaster if their selection and use is not carefully planned and implemented. For some types of tools, there can actually be a requirement for more time to develop, implement, and run a test set the first time than there would be if the test cases were executed manually. In order to choose the right tool, it is important that requirements be formulated for the selection and use of the tool. There may be no one tool (of a certain type) that fulfills all the requirements. Different potential users of the tool may also have different needs. Therefore, it may be possible that more than one of the same kind of tool may be selected, especially if there are many different environments in which the application being tested must run.

**Operational Specifications:** Test operation is the process of executing all or a selected number of test cases and observing the results. Although, preparation and planning for test operation occur throughout the software development life cycle, the operation itself typically occurs at or near the end of the software development life cycle. By-products of test operation are test incident report, test logs, testing status, and results, as illustrated in figure 7.2(d). According to Martin Pol, "test execution may only consume 40 percent or less of the testing effort," but it typically has to be concluded as quickly as possible, which means that there is often an intense burst of effort applied in the form of long hours and borrowed resources. Frequently, it also means that the test execution effort will be anxiously scrutinized by the developers, users, and management.

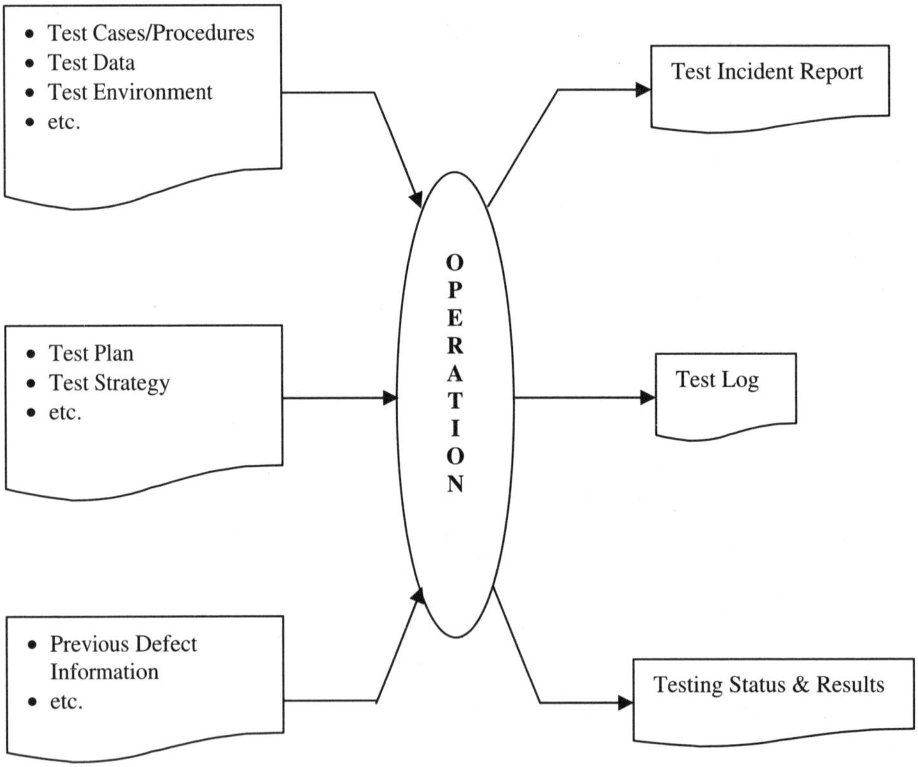


**Figure 7.2(d): Operational Specifications**

The results of each test case must be recorded. If the testing is automated, the tool will record both the input and the results. If the tests are manual, the results can be recorded right on the test case document. It may also be adequate to record whether the test case is passed or failed. Failed test cases will also result in an incident report being generated. *Test log* is a chronological record of relevant details about the operation of test cases. The purpose of test log is to share information among testers, users, developers, and others and to facilitate the replication of situation encountered during testing.

*Incidents* may be defined as any unusual result of executing a test. Incidents may be categorized as defects or enhancements, or merely retain their status as an incident if they are determined to be inconsequential or the result of a non-time anomaly. An incident report provides a formal mechanism for recording software incidents. *Testing status report* is often the primary formal communication channel that the test manager uses to inform the rest of the organization of the progress made by the testing team. The purpose of the test summary report is to summarize the results of the testing activities and to provide an evaluation based on these results. It allows the test managers to summarize the testing and to identify limitations of the software and the failure likelihood.

## 7.3 LEVELED TEST PLAN

Test planning is one of the keys to successful software testing, yet it's frequently omitted may largely be due to time constraints, lack of training or cultural bias. Testing without a plan is analogous to developing software without a project plan and generally occurs for the same reason-pressure to begin coding as soon as possible.

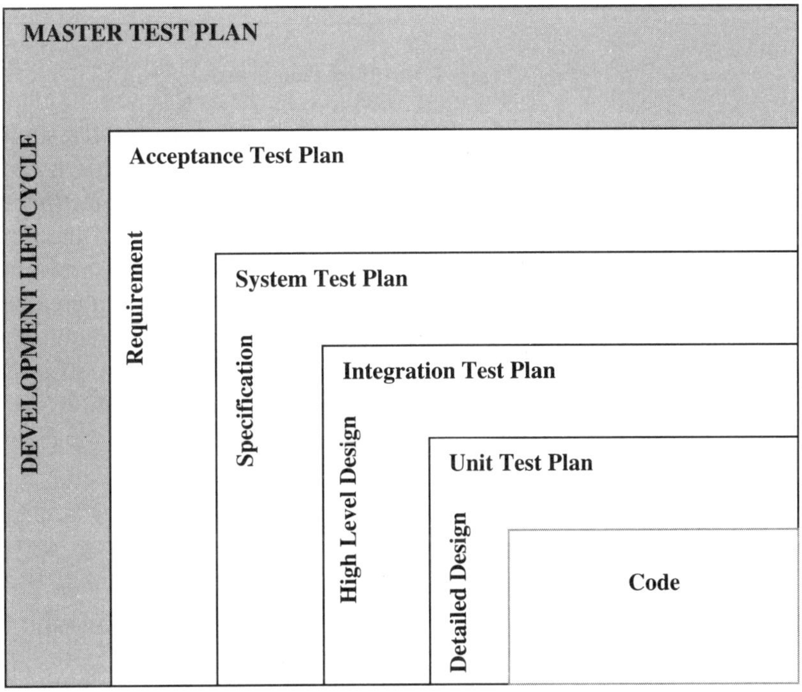


**Figure 7.3(a): Levels of Test Planning**

Test planning may occur at all the stages of software development life cycle. Integrated plan for all the stages is called Mater Test Plan (MTP), which may be a separate high-level document or could be included as part of the project plan. The purpose of MTP is to orchestrate testing at all levels as shown in figure 7.3(a).

Test planning is a process that ultimately leads to a document that allows all parties involved in the testing process to proactively decide what the important issues are in testing and how to deal with these issues. The goal of test planning is not only to create a long list of test cases, but rather to deal with the important issues of testing strategy, resource utilization, responsibilities, risks, and priorities. Discussing issues of what and how to test early in the development life cycle may save a lot of time, money and disagreements later.

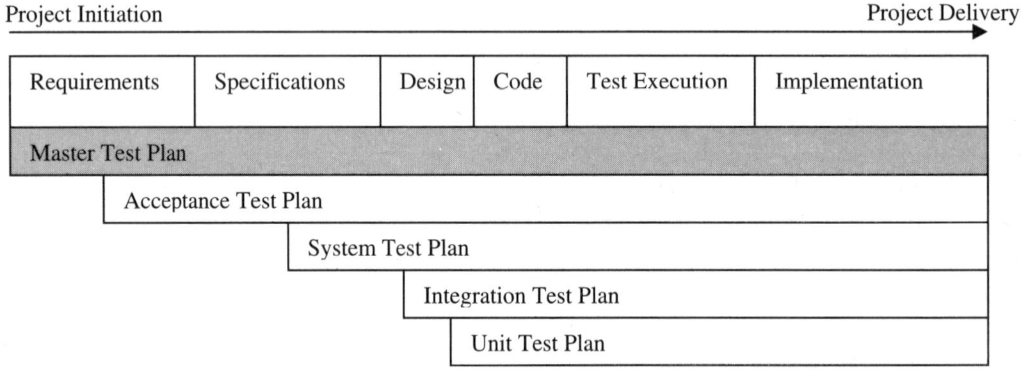


**Figure 7.3(b): Test Plan Schedule**

Test planning should be started a soon as possible. Master test plan should begin at about the same time the requirements specifications and the project plan are developed. The test schedule should be built around the milestones contained in the project plan such as delivery dates of various documents and modules, availability of resources, interfaces and so forth. Then, it will be necessary to add all of the testing milestones. These testing milestones will differ in level of details depending on the level of the test plan being created. In a master test plan, milestones will be built around major events including requirements and design reviews, code delivery, completion of user manuals, and availability of interfaces. In a Unit test plan, most of the milestones will be based on the completion of various software modules. Figure 7.3(b) shows approximate start times of various test plans with the software development life cycle.

## 7.4 DEVELOPMENT OF TEST PLAN

Effective test strategy and test plans are the foundation for successful testing. The test strategy is created when the application requirements are defined. The test plan provides the details that, when implemented, will meet the objectives outlined in the test strategy. The test plan is a living document that will change as the application functions become more clearly

defined and stable. There are two major purposes of planning for software testing. The first purpose includes identification of items to be tested, testing tasks to be performed, personnel responsible for each task and the risks associated with the plan. The second purpose of planning is to develop the scope, approach, resources and schedule of the testing activities.
The process of creating the test plan, along with the test plan itself, should be reviewed and agreed to by the members of the test team, the development team involved in the development, the end user and ultimately the management team. Following section describes the steps involved in planning for software testing.

### 7.4.1 Necessary Acquisitions

Test planners often get frustrated when they begin their planning process early and find out that all of the information needed is either not available or in a state of flux. This section of the test plan describes what is to be tested, and the pre-requisites to conduct the tests. Therefore, as a first step towards plan for testing is to check and verify the availability of the necessary acquisitions including the following.
**Test Items:** This section of the test plan describes programmatically what is to be tested within the scope of test plan, and verifies its availability. All the items necessary to perform the planned testing activities at the test site(s) should be identified by their names, numbers, and versions, as applicable. Following test items have been identified:

*Software Items*: Software items necessary to conduct the planned test activity includes operating systems, compilers, communications software, related applications software, databases, input file, code auditors, dynamic path analyzers, test drivers, preprocessors, test data generators, test control software, and other special test software. Availability of these items should be checked.

*Hardware Items:* Hardware items that will be used in the software test environment at the test site(s) should be identified and verified for their availability by their names, numbers, and versions, as applicable. The hardware items may include computer hardware, interfacing equipment, communications equipment, test data reduction equipment, apparatus such as extra peripherals (tape drives, printers, plotters), test message generators, test timing devices, test event records.

**Table 7.4.1: Testing Type and Responsible Groups**

| **Testing Type** | **Responsible Group** |
|---|---|
| Integration Testing | Testers |
| System Testing | Testers |
| Regression Testing | Testers |
| User Acceptance Testing | End-user Representatives |

*Other Materials:* Any other material needed for the testing at the test site(s) should be identified and checked. These materials may include manuals, software listings, and media containing the software to be tested, media containing data to be used in the tests, sample listings of outputs, and other forms or instructions.

*Participating Organizations*: This section will identify the organizations participating in the testing at the test sites(s). Responsibilities and roles of such organizations will also be included, by inserting the organization's chart. Each testing group consists of the project's test manager, and the tester(s). Table 7.4.1 shows the group responsible for the respective types of testing.

**Requirement Specification:** It is considered that requirements are the foundation stone on which an entire software product is built. Requirement specification document is a documentation of what a particular product or system should be or do. Requirements are used as inputs into the design phase of product development. Requirement phase is one of the major phases for the success of a software development life cycle (SDLC). In recent survey it is observed that about 71 percent of the software is not completed due to poor requirement specifications. Software Requirements define the required functional requirements (what the software must do), performance requirements (how many and how fast) and interface requirements (with, what, how, and with whom the software must interact). These requirements go through an iterative process, clarifying concepts and additional level of details at every iteration.

Software project frustration and failure cause due to slighting processes of requirement, development and management. Most of the software is not completed just because of negligence in the requirement phase. Earlier it was not the trend to appoint a requirement engineer whose main goal is to collect the complete and correct requirement from the stakeholders. It was the duty of any untrained person to just gather the information from the customer, and most of the software engineers are poorly trained to elicit, analyze, and specify the security requirement. They are often confused with the architectural security mechanism that is traditionally used. The few that have been trained have been only given an over view of security architectural mechanisms such as passwords and encryption rather than in actual are specified at all. Therefore, one of the major steps towards testing is to identify the complete requirement to be tested. This will help to derive the test cases using the current program specification.

**Design Specification:** It is required to specify refinements of the test approach and to identify the features to be tested. Test design specifications and its associated tests should be presented for this purpose. A test design specification may normally include the following structure:

- Test design specification identifiers;
- Approach refinements; and
- Test identification.

Test design specification identifier will specify the unique identifier assigned to this test design specification, while approach refinements will specify refinements to the approach described in the test plan, and include specific test techniques to be used. The method of analyzing test results will be identified through approach refinements. Test identification lists the identifier and a brief description of each test case as well as the procedure associated with this design. A particular test case may be identified in more than one test design specification.

**Test Case Specification:** Test case specification defines a test case identified by a test design specification. The test case specifications should be developed from the test plan and are the second phase of the test development life cycle. The test specification explains how to implement the test cases described in the test plan. A test case identified may be referenced by test design specifications used by different groups latter on. Therefore enough information should be included in test case specification. This will increase the reusability of the test cases generated. A test case specification may normally include the following structure:

- Identifier;
- Test items;
- Input and output specifications; and
- Special procedural requirements.

Test case specification identifier will specify a unique identifier assigned to this test case specification. Another important building block to test case specification is test items that need to be identified and briefly described. Along with each test items, following documentations should be provided:

- Requirements specification;
- Design specification;
- Users guide;
- Operations guide; and
- Installation guide.

Each input required to execute the test case, and the relationships between the identified inputs should be specified. Similarly, it is required to specify all the outputs and features required of the test items as well as the exact value for each required output. There may be any special constraints on the test procedures, like special set up, operator intervention or output determination procedures that execute the test case. These special constraints should be described in detail.

**Test Schedules:** In this section, the test milestones should be identified in the software project schedule as well as all item transmittal events. One of the major objectives of test schedule includes identifying the testing planning details within the project planning schedule. This requires the preliminary schedule of the test effort and needs to be updated from information gathered during test case development.

The test schedule controls all activities during the test effort. It also provides management with an explicit representation of all test activities. It includes and defines any additional test milestones if needed. Time required to do each testing task will be estimated. For each testing task and test milestone, schedule will be specified. Period of each testing resource utilized during testing will be specified. The section contains the overall project schedule. It discusses the phases and key milestones as they relate to quality assurance. It discusses the testing goals and standards that are likely to be achieved for each the phases of testing that will be deployed, e.g., Usability Testing, Code Complete Acceptance, Beta Testing, Integration Testing, Regression Testing, and System Testing.

**Test Deliverables:** This section will identify the deliverable documents after executing the test. The major deliverables are testing input data, test output data and test tools, including module drivers and stubs. The following documents should be included:

- Test plan documents;
- Test design specifications;
- Test case specifications;
- Test procedure specifications;
- Test item transmittal reports;
- Test logs;
- Test incident reports; and
- Test summary reports.

Test plan document should address testing objectives, criteria, standards, schedule and assignments, and testing tools.

### 7.4.2 Testability Assessment

Testability has been defined as one of the attributes of software that bare on the effort needed to validate the software product. Several characteristics have been identified that contributes to a software system's testability. There is a range of typical factors that are expected to facilitate testability in assessing the *Level of Testability* during testing including *Internal*, *External*, and *Environmental Facilitators.*

Internal facilitator including operability, controllability, simplicity, stability, understandability, observability, etc. of software assists to verify how easily a program can be tested. If all the internal facilitators are available for the software under test, the software will be *L-1 Testable.*

One of the important facets of testability that impacts the ease of testing is the amount of difficulty in setting up drivers to execute a component and creating stubs for functionality that does not yet exist. The external facilitators including documentation, coding standard, test suits, data, etc. will help in verifying whether the software is of *L-2 Testable.*

Environmental facilitators may ultimately decide software system to be highly testable, and hence considered to be a major contributor for assessing software testability. Environmental facilitators include test tools, testing criteria, timing issues, implementation, testing skills, etc. Environmental facilitators will verify for the software system to be *L-1 Testable*.

## 7.5 MASTER TEST PLAN

Master test plan may be considered as one of the major communication channels with all project participants. Ideally, it is desirable to begin the Master Test Plan at about the same time the requirements specifications and project plan are being developed. Master Test Plan is also called high-level test plan, as it is prepared on the basis of high-level documentation instead of detailed information. The Master Test Plan should address how changes to the plan itself and its corresponding detailed test plans will be handled. Each development phase within the software development process creates a work product that can be tested to see how successful the translation is. Contingency planning minimizes the disruption when a risk is realized. The team does not need to spend effort determining 'what to do?', and the cycle time to receive an authorization to proceed should be minimal. The test team, the application development team, all user groups and management review and approve the Master Test Plan.

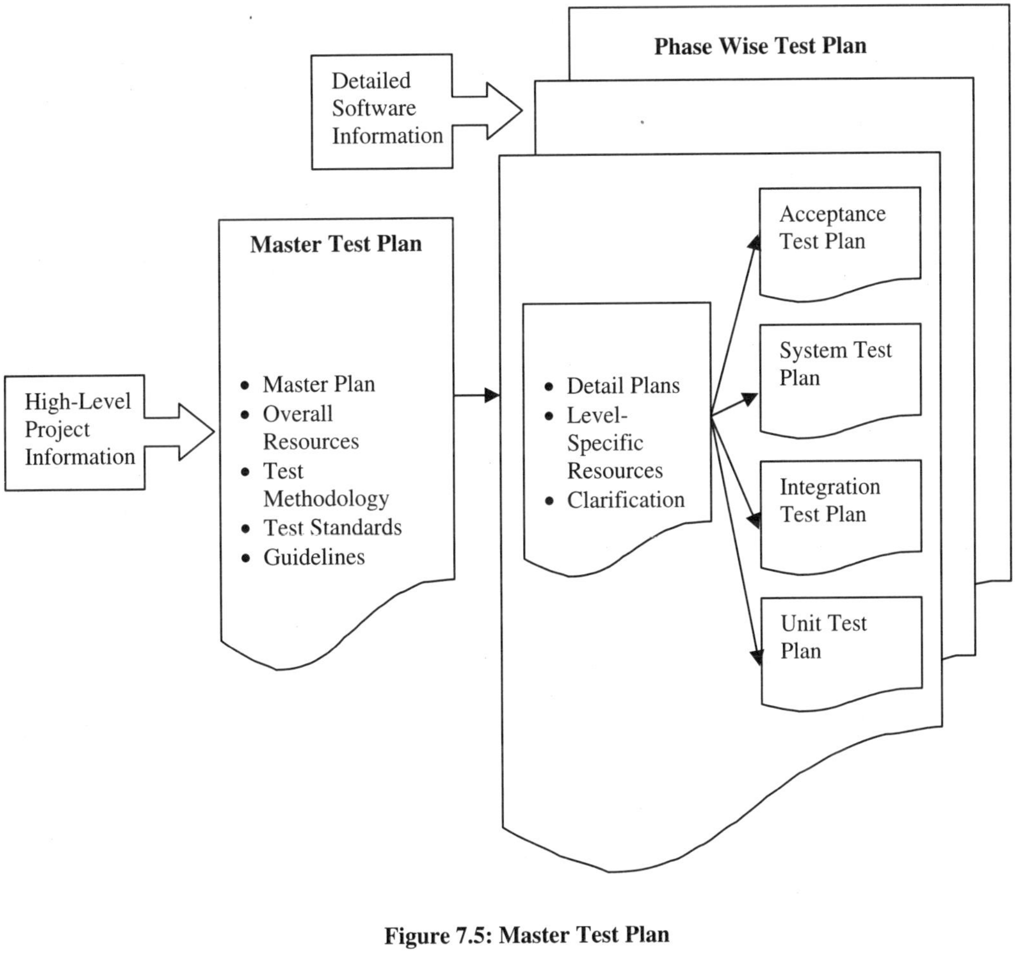


**Figure 7.5: Master Test Plan**

To be most effective, tests should be conducted early to prevent defect migration. In order to conduct testing early in the development life cycle, test planning must start at the beginning and proceed in parallel with the development process. General high-level information is used to develop the Master Test Plan. On the other hand, more specific detailed software information is used to develop phase wise detailed test plan as shown in figure 7.5.

## 7.6 PHASE WISE TEST PLAN

Test plan is developed at the time application requirements are defined, which provides the team adequate time to define the test, locate and configure the hardware and software resources, locate and train the human resources and schedule the tests. Figure 7.6 depicts how the realism of the test environment increases at higher levels of test. Test planning at various phases has been discussed briefly in the following scctions.

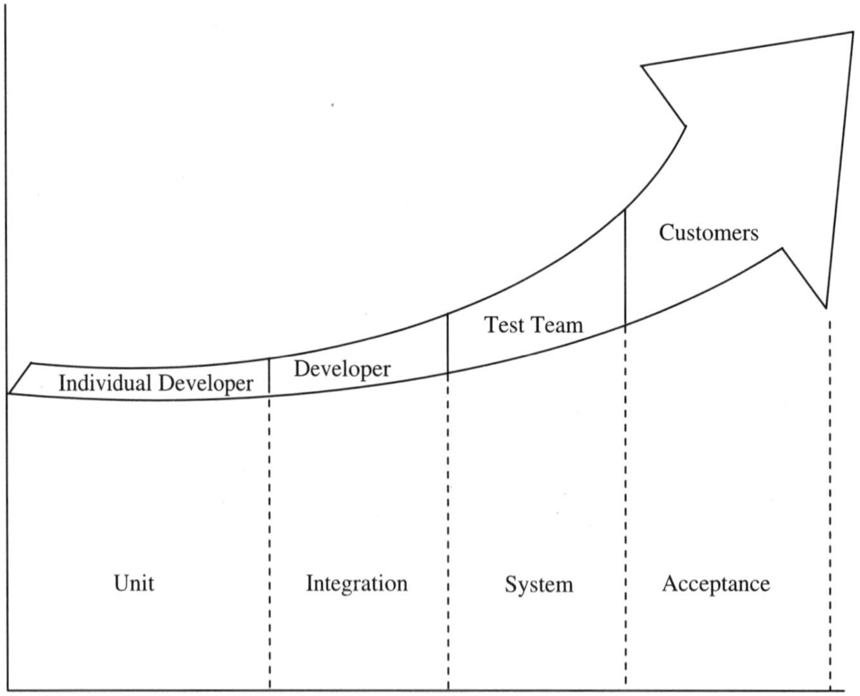


**Figure 7.6: Typical Test Levels**

### 7.6.1 Acceptance Test Plan

Acceptance testing is based on the user's requirements and demonstrates whether the intended goals have been achieved. It is planned first as it participates in modeling the system when it is complete. Acceptance testing is most effective testing when performed by users or their representatives in most realistic environment possible. Acceptance test is not intended to be comprehensive, and therefore does not last as long as the systems test.

Planning for acceptance testing should begin as soon as the high-level requirements are available, that is, early in the life cycle. One of the major objectives of acceptance test plan is to develop the criteria for what constitutes a completed project which will provide the basis for acceptance of the product. Acceptance test plan and developed test cases during planning should describe the behavior and nature of the product. Project plan, master test plan, and requirement documentation forms the basis for writing acceptance test plan.

Acceptance test cases are based on requirement specifications. User documentation may be very useful when planning a new release of an existing system. At all levels of test, enhancement requests, and defect reports are treated as request documents. The planning for acceptance testing is the responsibility of the test team leader and should be developed with reference to the overall development and test plan. It is assumed that acceptance testing will take place in the live environment; the test team leader must liaise with the system administrator to plan installation of the application under test prior to testing and to ensure that the administrator is aware of the date, time and duration of the system test.

### 7.6.2 System Test Plan

System test may go on for an extended period of time and is intended to be as comprehensive as the resources allow and risks demand so. During system testing, load testing, performance testing, reliability testing and other typical testings are performed. The system test plan can normally be started as soon as the first draft of the request is fairly complete. The software design documentation will also have an impact on the system test plan. Requirement traceability matrix built in acceptance testing may be taken as input to system test plan. It also helps in constructing system traceability matrix.

System test plan is in agreement with the master test plan and seek to amplify those parts of the master test plan that addresses system testing. It is based on request specifications, design documentation, and user documentation. The planning of system test planning is the responsibility of the test team leader and should be developed with reference to the overall development and testing plan for the application under test. It is assumed that system testing will take place either in a dedicated test environment or in the live environment depending on the availability of the suitability specified test environment. It can also be assured by an assessment of the commercial or safety critical nature of the live environment and the likelihood of the system test adversely affecting it. It is essential that the system test plan take account of time and effort required for correction of defects and retesting.

### 7.6.3 Integration Test Plan

Integration test is done to ensure that the various components of a system interact and pass data correctly among one another. It may be accomplished at various levels including lowest level where developers ensure whether units work properly together and higher level done by the test team. One of the key issues in integration testing is to determine which model is created and integrated first. Integration tests plan are built based upon the detailed design specifications. Many strategic issues are there to be considered during planning for integration testing including the following:

- Identification of modules or objects to be assembled and tested as a group;
- Description of functional subassemblies;
- Estimation of testing appropriateness; and
- Verification of implementation based testing objectives.

The planning of integration test is the responsibility of development team leader and should be conducted with reference to the overall development plan fort application under test. It is assumed that integration testing will not take place in the live environment but in the development environment or in separate test environment.

### 7.6.4 Unit Test Plan

The objective of unit testing is to ensure that reliable program units are produced that meet their requirements. The planning for unit testing is the responsibility of development team leader and is developed with reference to the overall development and testing plan for the application under test. It is generally assumed that unit test planning will not take place in the live environment but in the development environment or in separate testing environment. It is essential that unit test plan take account of the time and effort required for any correction of defects and retesting.

## 7.7 CONCLUSION

Software test planning is the most important phase which provides an opportunity for the tester to determine what to test and how to test it. Test planning determines the scope, approach, resources, and schedule of the intended testing activities. Test plan generally guides all testing activities, and a substantial investment is required for best planning, willingness to revise the plan dynamically to account for changes in requirements, designs, or code as bugs are uncovered, to achieve efficient testing. Test plan also identifies the items to be tested, the features to be tested, and the type of testing to be performed. Test-design specification, test-case specification, and test-procedure specification are the three types of document in test specification. Testing should run in parallel with the development life cycle.

Test strategy outlines the steps in the overall testing process, describes the objectives and identifies the testing levels. Testing strategy also describes the overall approach for testing and verification. Testing completion criteria for the overall test plan needs to be set in advance. Test design specification identifier will specify the unique identifier assigned to the test design specification, while test specification explains how to implement the test cases described in the test plan. Test design specifications used by different groups may refer the test case identified latter on.

## KEY TERMS

Software Test Plan, Plan Format, Test-Design Specification, Test-Case Specification, Test Procedure Specification, Test Process, Test Strategy, Readiness Criteria, Pass/Fail Criteria, Completion Criteria, Acceptance Criteria, Software Test Environment, Software Items, Hardware Items, Test Schedule, Risk, Test Deliverables, Test Report

## REVIEW QUESTIONS

## OBJECTIVE TYPE QUESTIONS

1. Software test plan is a________
   (a) Managerial document
   (b) Technical document
   (c) Both managerial and technical document
   (d) None of the above
2. Test planning should take into account____________
   (a) Static testing
   (b) Dynamic testing
   (c) Both static and dynamic testing
   (d) None
3. _____________ describes the overall approach and techniques to be used in each phase during testing.
   (a) Test planning
   (b) Test strategy
   (c) Both test plan and strategy
   (d) None of the above
4. The information about the test data, test cases, and test configuration used during testing process will be submitted via_______
   (a) Unit Test report
   (b) Bug Report
   (c) System Report
   (d) Unit/System report

5. The completion criteria for overall test plan should be set__________
   (a) In advance
   (b) Before testing
   (c) In parallel with development process
   (d) None of the above

## SHORT ANSWER TYPE QUESTIONS

1. In which phase of software testing process detailed testing scripts need to be written?
2. How is acquisition phase different from execution phase of software test process?
3. What do you understand by test planning? Explain in brief.
4. Describe the need and significance of software test plan.
5. List the role of a test plan and elaborate its benefits.
6. What is the purpose of revising test plan dynamically?
7. Identify appropriate target audience of a test plan?
8. How does software test plan map tests to requirement?
9. What is a software test plan format and how can you develop it?
10. Describe the importance of documentation during testing.
11. What do you understand by bug reporting and how can it be generated?
12. Describe the information documented in system test report.
13. Describe the basic criteria for successful testing.
14. What is the completion criterion and how can you check the same?
15. Identify the importance of test schedule during preparing test planning.
16. What is the significance of including test cases and specification in test plan?
17. List all the documents provided with each test items.
18. How can you generate a test report? Explain in brief.
19. List the deliverables at the end of the testing process.

## DESCRIPTIVE QUESTIONS

1. Efficient testing requires a substantial investment in planning. Justify.
2. Propose an entry and exit criteria for each phase of the testing processes.
3. How can a software test plan help in assessment of test-related risks? List a plan for their mitigation?

4. Elaborate testing process in detail. Take a very small project, follow the process discussed and test the software.
5. Prepare a complete test plan for software project, gradually while development of the software is going on.

## REFERENCES


1. Pressman, Roger S. Software Engineering - A Practitioner's Approach. Fifth edition. The McGraw-Hill companies, Inc.
2. Kaner, C., Falk, J., Nguyen, H.-Q. Testing Computer Software. Wiley Computer Publishing, 1999.
3. IEEE Standard for Software Test Documentation IEEE Std 829-1983
4. IEEE Standard for Software Test Documentation IEEE Std 829-1998 Approved 16 September 1998
5. John Watkins, Testing IT- An Off-the Shelf Software Testing Process, Cambridge University Press, 2001.
6. Rick D Craig & Stefan P Jaskiel, Systematic Software Testing, Artech House Publishers, 2005.


## USEFUL LINKS

www.acutest.co.uk/testing-service.html

www.tbs-sct.gc.ca/emf-cag/acceptance/outline/atpo-vper_e.asp

www.toa.com/shnn?toos

www.usdoj.gov/jmd/irm/lifecycle/appendixc15.htm

cit3.cdn.swin.edu.au/majorprojects_2004/js/documents/CATP_R2.doc

www.cmswatch.com/Feature/117-Testing

http://www.ing.iac.es/~eng/standards/software/sof-std-4/node9.html

## Chapter 8

# Object-Oriented Testing

*I didn't fail the test, I just found 100 ways to do it wrong.*
**Benjamin Franklin**

### OBJECTIVES

A successful completion of this chapter will enable the reader for the following:

- To understand object-oriented software development process.
- To realize the need and importance of integrating testing within development process.
- To appreciate the need and importance of testing in quality estimation.
- To compare traditional approach with object-oriented software development approach.
- To discuss object-oriented paradigm.
- To elaborate the process of testing object-oriented software.
- To define pertinent issues in testing object-oriented software.
- To describe objects and classes.
- To understand characteristics of object-oriented software.
- To realize the fact related class-object life cycle.
- To elaborate the impact of object orientation on testing.
- To describe object-oriented software testing models.
- To discuss object-oriented software testing strategy.
- To understand object-oriented requirement testing.
- To describe object-oriented design testing.
- To elaborate testing strategy involved in object-oriented design testing.
- To discuss unit testing and elaborate integration testing.
- To apply the relevant techniques in object-oriented software testing.

## 8.1 INTRODUCTION

Irrespective of the technology employed in the development of a software system, any testing process will involve a number of common activities that must be performed by testers with appropriate skills and completed in a particular order to ensure successful testing. This is the essence of a repeatable, predictable and reusable process. Object orientation is one of the many approaches available for developing software applications. It is fundamentally different from procedural programming in the way it views process and data. In writing object-oriented applications, developers are encouraged to reuse previously created objects in order to save time, effort, and cost. This approach can also improve quality of the delivered software by allowing developers to reuse tried and tested objects.

It is widely accepted that the object-oriented paradigm is much more natural than the earlier prevailing paradigm. It is likely to significantly increase software reusability, extendibility, interoperability, and reliability. Object-oriented software is developed in iterative as well as recursive increments. It has a different structure and behavior than software implemented in traditional procedural languages. Object-oriented software is not functionally decomposed into separate procedures rather it is composed of objects and classes. These objects and classes interact via message passing and usually share definitions through inheritance. Moreover, it is not developed with the traditional waterfall life cycle, rather, it is developed incrementally with iterative and recursive cycle including analysis, design, implementation and testing phases.

Software testing began to see its importance from the late 1950. Recent developments have also demonstrated its significance, because of both experience and economics. Software testing is one of the most critical elements of software quality assurance. The importance of software testing and its implications with respect to software quality cannot be overemphasized. It is also responsible for bug prevention. The pragmatic goal of software testing is to discover bugs, symptoms caused by bugs, and provide clear diagnoses so that bugs can be easily corrected. Bugs are not always obvious. Different bugs can have the same manifestations and one bug can have many symptoms. The symptoms and causes can be disentangled only by more tests, small and detailed tests. The fundamental difference discussed above has a large impact on how object-oriented software is tested.

The differences between 'old' and 'new' ways of testing software are basically focused on objects instead on functions that transform inputs to outputs. The most significant difference is in the way object-oriented software is designed as a set of objects that essentially model a problem and then collaborate to affect a solution. Underlying this approach is the concept that while a solution to a problem might need to change over time, the structure and components of the problem itself does not relatively change so much or as frequently.

Object orientation in combination with iterative development techniques, rapid prototyping and object reuse, provide a massive challenge to quality assurance people in terms of significantly larger testing load as compared with more traditional development techniques. This increased testing load presents the biggest impact of object orientation on the management and planning of testing. Those aspects of the testing process that employ Black Box testing techniques are largely unaffected by the use of object-oriented programming because test cases designed using such techniques are constructed with no knowledge of the internal structure of the application under test or its components. Automated testing tools should also be considered to improve reuse in testing object-oriented software, as well as to save testing and effort and increase confidence in the quality of software.

## 8.2 OBJECT-ORIENTED PARADIGM

Object-oriented paradigm is one of the most prominent and grown up software development paradigms now. While the procedural paradigm is task-based that focuses on the actions the software. Object-oriented paradigm is both task-based and data based. Hence the object-oriented paradigm combines together relevant data and tasks in entities known as objects. The object-oriented paradigm is centered on the concept of the object. Everything is focused on objects. Objects have traditionally been defined as data with methods. Unfortunately, this is a very limited view of looking at objects. Objects inherently have knowledge about itself. Data in an object allows it to know what state it is in and the code in the object allows it to operate upon them properly. An object inherits all of the attributes and operations from its parent class, in addition to having its own attributes and operations.

An object can also become a class for other objects, forming another branch in the hierarchical tree structure. When an object is applied and contains data or information, it is an instantiation of the object. Objects interact or communicate by passing messages. When certain messages are passed between two objects, the objects are coupled. A class is a definition of the behavior of an object. It contains a complete description of the data elements the object contains, the methods the object can do, and the way these data elements and methods can be accessed. The object-oriented design paradigm suggests that natural objects (that is, the entities) occurring in a problem should first be identified and then implemented. A simple way of identifying the functions performed by a system is by examining the action verbs occurring in the problem description, since verbs represent activities performed by a system.

The object-oriented paradigm proposes the use of hierarchically related abstract data types to implement a system as a collection of highly independent subsystems. These interact with one another through small, carefully designed interfaces. Each subsystem is unable to access implementation details of other subsystems. The management of the desired level of visibility in this way is called 'information hiding' and is said to yield designs, which have a lower degree of coupling between different subsystems. Lower coupling enables to translate into greater ability to implement separate systems independently. The stronger the coupling between software artifacts, (i) the more difficult it is to understand individual artifacts, and hence to correctly maintain or enhance them; (ii) the larger the sensitivity of unexpected changes and defect propagation effects across artifacts; and (iii) consequently, more testing efforts are required to achieve satisfactory reliability levels. Table 8.2 depicts a comparison between procedural vs object-oriented paradigm.

Object-oriented paradigm provides a uniform structure for all the components in the form of a procedural interface. Classes provide an excellent structuring mechanism. They allow a system to be divided into well-defined units, which may then be implemented separately. Classes support information hiding. A class can export a purely procedural interface and the internal structure of data may be hidden. This allows the structure to be changed without affecting users of the class, thus simplifying maintenance.

**Table 8.2: Procedural Vs Object-oriented Paradigm**

| Procedural Paradigm | Object-oriented Paradigm |
|---|---|
| • Programs are written in a procedural fashion.<br>• Statements are written in the form of a batch.<br>• It is quite difficult to debug the larger programs.<br>• Functions are used.<br>• Data is treated as a stepson and functions were given more priority.<br>• The original data could easily be accessible to all the functions even to those, which do not have permission to access, and hence get corrupted. | • Data is given higher priority.<br>• Data and functions are packed into one box known as an object.<br>• Unauthorized access of the data is prevented.<br>• Programmer can define the own data type and the corresponding functions.<br>• Objects may be reused by other program, and hence increases reusability of the program.<br>• Much reliable than the Functional Paradigm.<br>• The paradigm provides the facilities of data hiding, reusability by Inheritance and by Polymorphism. |

Object-orientation encourages and supports software reuse. This may be achieved either through the simple reuse of a class in a library or via inheritance, whereby a new class may be created as an extension of an existing one. In both cases the result is a reduction in the amount of software, which must be written and, as a result, an improvement in the reliability of the resultant system since previously tested classes may be utilized.

There are reported misconceptions regarding the process of testing. Practitioners generally argue that testing will not be difficult (or even necessary at all) if a proper design and implementation process has taken place. Unfortunately or fortunately, the advantages of object-orientation become potential disadvantages when testing is considered. The structuring of the system as a set of independent classes requires that each of these must be tested and there may be a large number of them. In addition, information hiding, which encourages designers of classes to have purely procedural interfaces, makes it difficult to determine whether the class is working correctly, since the state of internal data may not be accessible via the interface. In order to test a class the programmer must be able undertake the following activities:

- Create an instance of the class, that is, an object, passing the appropriate parameters to the constructor.
- Call the methods of the object passing parameters and receiving results.
- Examine the internal data of the object.

Moreover, there are several fundamental concepts and characteristics with regard to this paradigm. These foundations of the object-oriented paradigm are described in brief in the following sections.

### 8.2.1 Objects

Objects are one of the basic building blocks while object-oriented methods are being used to analyze or design a complex software system. Figure 8.3(a) depicts a generic view of Objects. In everyday life, an object is anything that is identifiably a single material item. It may be a car, a house, a book, a document, or a paycheck. In everyday life, an object is anything that is identifiably a single self composed material item. An object can be a car, a house, a book, a document, or a paycheck. This concept of defining an object may be extended a bit in terms of anything that is a single item to represent in a program. Hence, living objects, such as a person, an employee, or a customer, as well as more abstract objects, such as a company, a database, or a country may also be included. Thinking about objects in this way not only enables to write code that models the real world; it also enables to break up a large program into smaller, more manageable units.

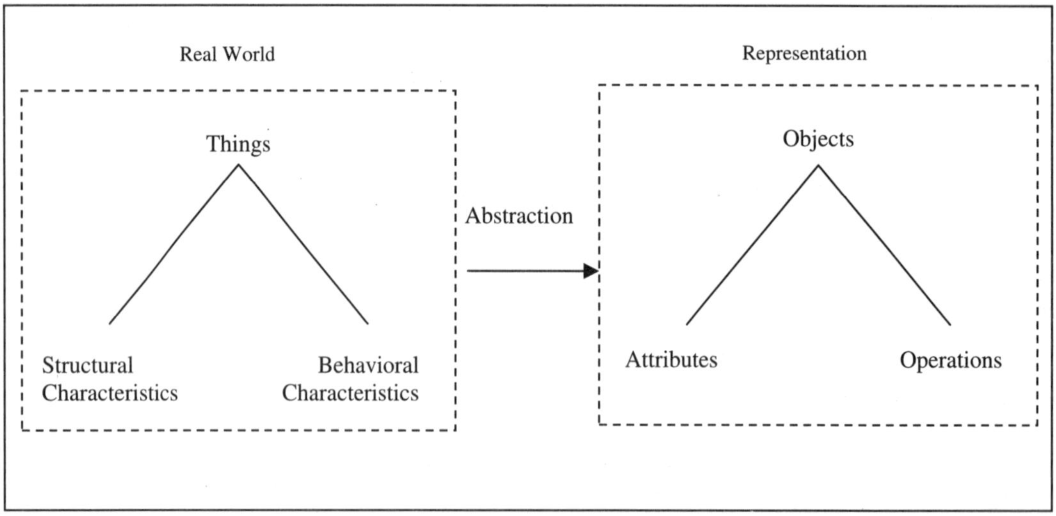


**Figure 8.2.1: Object Representation**

The idea comes from the concept of a *black box* where a lot of objects are available to be used but of which the mechanism is not understood. Black boxes formalize the idea that there's a difference between what something does and how it works, and that two objects can do the same thing but work differently inside. Replacing one object with another does have some subtle effects. An object is something that has crisply defined boundaries, and may be defined as something that has state, behavior, and identity. The structure and behavior of similar objects are defined in their common class; the terms instance and objects are interchangeable. Objects are well-defined representational constructs of concrete or conceptual entities. It encapsulates various structural characteristics called *attributes*. These attributes may determine the possible states of an object, and are extracted from a domain using declarative or data-oriented abstraction. These attributes may be single-valued or multi-valued. Single-valued attributes resolve to one value, while multi-value attributes resolve to a collection or set of values. Figure 8.2.1 depicts a generic view of object representation.

Objects also encapsulate behavioral characteristics called *operation*. An operation is representational constructs of behavioral characteristics of entities, and determines the possible behavior of an object. It has a signature consisting of a name, input parameters, output parameters, and possibly return parameters. Objects may be active or passive. An active object has a thread of control or may initiate activity. Active objects may take the initiative to request services from other objects. On the other hand, passive objects do not have a thread of control and may not take the initiative to request services from other objects unless they receive control from an active object. Objects may be persistent or transient. Persistent objects exist after their creator has ceased to exist. Transient objects exist only during the time that their creator exists.

All objects are unique and distinguishable from other objects. Objects may be compared to verify for their pure identity. This will yield true for same objects. They may also be compared for shallow equality, which yields true for objects with the same attribute values. Objects may also be compared for deep equality yielding true for objects with the same attribute values recursively within their subordinate parts.

### 8.2.2 Classes

It is apparent that there are two sides to an object including what it does, which is usually publicly known, and how it works, which is usually hidden. In programming, the 'what it does' is normally represented in the first instance by *methods*, which are blocks of functionality used. Secondly 'how it works' is represented both by methods and by any data (variables) that the object stores. In general, a class is defined by its fields and methods. In everyday terms, a class may be defined as a group, set or kind marked by common attributes or a common attribute; a group division, distinction, or rating based on quality, degree of competence, or condition.

Class may be defined in the context of object-oriented analysis and design as a set of objects that share a common structure and a common behavior. A single object is simply an instance of a class. A class is concerned with the implementation of uniform structural characteristics and behavioral characteristics. A class has an intentional notion or capability to define a pattern for objects of the class. It also has an extensional notion or capability to create objects of the class. The class is known as an object factory. Extensional capability of a class maintains references to all objects of the class. Figure 8.2.2(a) depicts a generic view of class representation.

Class encapsulates structural characteristics called attributes, which are implemented by the class, and may be associated with objects of the class, or with the class as a whole. Class also encapsulates behavioral characteristics called operations, which are implemented by the class as methods or subprograms, and are known as services offered by the class. These operations may be associated with object of a class, or with the class as a whole. Class may define the generally three accessibility criteria of attributes and operations from outside an object of a class including public accessibility, protected accessibility and private accessibility. A class may be parameterized, i.e. it may require parameters in order to become a defined class that is capable of installation. The parameters are bound to actual values to create a real class. Parameterized classes are known as *template classes* or *generic classes*.

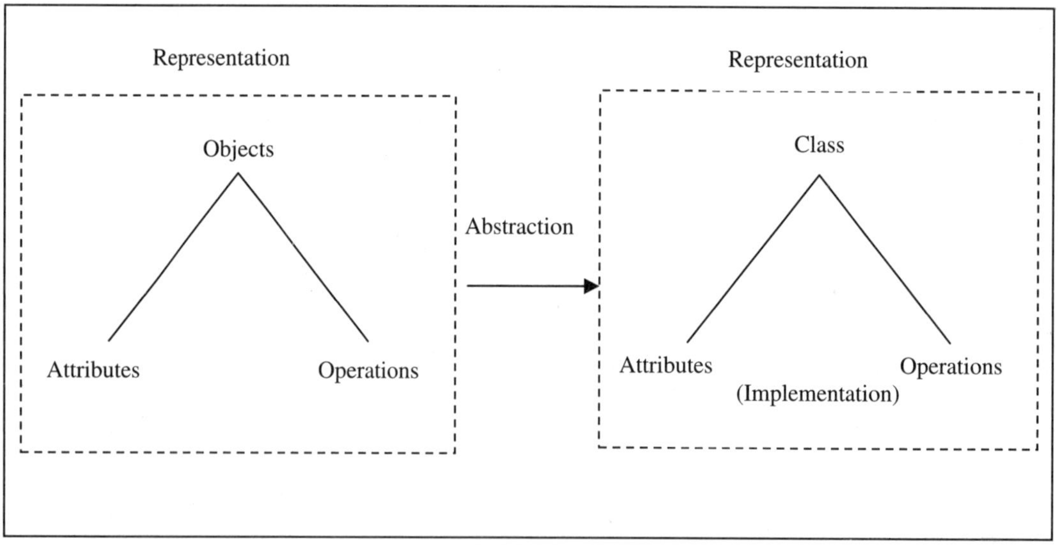


**Figure 8.2.2(a): Class Representation**

Classes may allow their objects to be classified statically or dynamically. Static classification disallows objects from changing their classes, while dynamic allocation allows it. Classes have a life cycle that is shared with their objects. Figure 8.3(c) depicts the life cycle of classes and objects with the operations including following:

- Initialization operations that are the class operations utilized to initialize a class, generally invoked when first object of the class is created;

- Contractor operations that create or construct objects of the associated class;
- Selector operations that provide an interface for getting information about and from within an object or class;
- Mutator operations which provide an interface for setting information about and within an object or class;
- Destructor operation that destroy objects of the associated class; and
- Finalizer operations that are implicitly utilized to uninitialize a class, generally invoked when the last object of the class is destroyed.

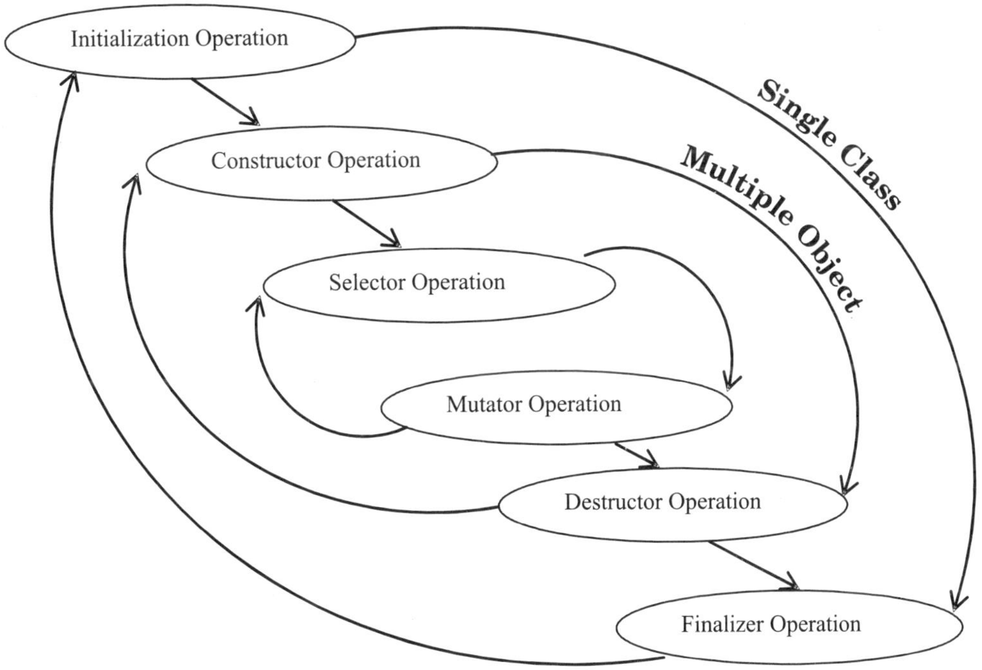


**Figure8.2.2 (b): Class-Object Life Cycle**

### 8.2.3 Characteristics

There has been phenomenal growth in the utilization of object-oriented technology for developing software systems during this decade. Object-oriented methods have proliferated in software applications for such diverse fields as artificial intelligence, graphics, exploratory programming, physics, telecommunication, banking, stock market, manufacturing, and the internet. The term object-oriented is used differently by different people. However, they all agree on its promising features for proving the quality of a software system. The popularity of object-oriented philosophy has been due to its claimed support for data abstraction, information hiding, extensional programming, and in particular, reusable software. Another major advantage of using object-oriented programming is its flexibility. This feature of object-oriented programming allows a software designer to produce reusable, intanciable, and information-hiding modules, also known as classes. Three fundamental characteristics required for an object-oriented paradigm are Encapsulation, Polymorphism and Inheritance. Polymorphism and Inheritance are two aspects unique to object-oriented approach, while encapsulation in not.

**Encapsulation:** Encapsulation has often been described simply as hiding data. Objects generally do not expose their internal data members to the outside world. Information hiding is a way of designing such that only the subset of the module's properties, its public interfaces, is known to users of the module. It gives rise to encapsulation in object-oriented language. Encapsulation means that all that is seen of an object is its interface, namely the operations we can perform on the object. Information hiding is a theoretical technique that has indisputably proved its value in practice.

**Polymorphism:** For an object-oriented system, polymorphism allows the implementation of given operations to be dependent on the object that contains the operations; an operation can be implemented in different ways in different classes. Polymorphism is an important concept that allows the building of a flexible system. This concept of polymorphism allows the developer to specify what shall occur and how. Polymorphism means having the ability to take several forms. Polymorphism assists the programmer to reduce complexity [52]. Definition of new classes of objects that inherit much of their behavior from previously defined classes is provided by this object-oriented paradigm.

**Inheritance:** Inheritance is a form of reuse that allows a programmer to define objects incrementally by reusing previously defined objects as the basis for new objects. Inheritance decreases complexity by reducing the number of operations and operators, but this abstraction of objects can be make maintenance and design difficult.

## 8.3 IMPACT OF OBJECT ORIENTATION

Object-oriented software is radically different from its procedural counterpart. Procedural programming relies on procedural autonomy with no interaction through nonlocal variables. On the other hand, object-oriented programming discourages procedure autonomy and relies on the packaging of procedures that share an object's local variables. Object-oriented technique provides an abstract way of thinking about a problem in terms of problem-domain concept, rather than in computer concepts. Object-oriented languages provide many new constructs that aid in the implementation of an object-oriented design. Since these concepts are not prevalent in traditional software, existing software testing techniques fail to suffice for completely testing object-oriented software.

Migration of an organization from traditional software development to object-oriented software development greatly impacts its culture, leading to the change in standards and procedures. Since testing of the software in an organization is also based on the traditional software philosophy, the organization must determine the impact of this technology transition on its testing techniques. Following section describes briefly the impacts of object-oriented characteristics on testing.

**Encapsulation:** The biggest impact of encapsulation on the testing of object-oriented software is the change in the definition of a unit test. The traditional approach of software testing is based on the assumption that the software is developed using top-down approach. The basic units of testing are subprogram or subroutine in traditional approach. While with respect to object orientation, subprograms are usually not a standalone entity. It is associated with an objects and their behavior. An object encapsulates its state and the associated functions. Thus, the basic unit of object-oriented software testing is an object or a class.

**Abstraction:** Abstraction separates the essential behavior of an object from its implementation, and presents on the outside view of an object. The information abstracted by the object is required to satisfactorily test that object. Therefore, the abstraction of the information impacts the testing strategy for object-oriented software.

**Inheritance:** Inheritance defines a relationship among classes, wherein one class shares the structure or behavior defined in one or more classes. If an object acquires characteristics directly from only one object, it is called single inheritance. On the other hand, if the characteristics are acquired directly from more than one object, then it is termed as multiple inheritances. Any strategy for testing object-oriented software has to accommodate the different types of inheritance relationships. So the testing strategy must consider the possibility of substitution of a base class where a derived class has been used.

## 8.4 PERTINENT ISSUES

No doubt, the development of a large software system is a time and resource-consuming process. The inception of object-oriented technology in the software industry has created new challenges. Inheritance, information hiding, encapsulation, and specially polymorphism, makes object-oriented software testing a bit more difficult than the traditional software testing. Software testing is an important software quality assurance activity to ensure that the benefits of object-oriented programming will be realized.

Object-oriented software testing has to deal with new problems introduced by the powerful new features of object-oriented languages. These features result in the so-called 'delocalized plan' in which several member functions from possibly several object classes are invoked to achieve an intended functionality. Often a member function of a class may in turn invoke other member function, resulting in the so-called invocation chain of member function.

Testing and maintenance of object-oriented software is costly because of the complex relationships that exist in object-oriented software. Few observations in this regard are listed as follows:

- It is generally difficult to understand a class in a large object-oriented program if that class depends on many other classes.

- Whenever a subclass or superclass is added to a class, there is a need to test the inherited methods in this newly formed context.
- A tester may not know where to start testing an object-oriented library without sufficient insight.
- Even if the overriding and overridden methods are semantically similar, there is a need to retest the classes in the context of overriding and overridden methods.
- Generally during testing of object-oriented software, tester has to understand the called function, possibly create and initiate certain objects and write code to simulate the behavior and effects of called functions. Therefore, it is extremely costly to construct test stubs in case of object-oriented software.
- If the order of specification of superclasses of a subclass is changed, the subclass must be retested even though only syntactic changes are made.

Object-oriented software is developed in iterative increments. After each increment the software is put together. Now object-oriented integration testing plays it's role, which tests the interactions and interfaces of the compounds that are put together. Integration testing assumes that unit testing has been done on the components and that the defects have been removed. These components are integrated during integration testing. Therefore the objective of the integration testing is to find the defects while these fault free components after unit testing integrate with each other in an inappropriate way. The complex nature of object-oriented software leads to many possible test cases and hence high cost of testing.

## 8.5 OBJECT-ORIENTED TESTING MODEL

Testing is one of the areas in software engineering where there is a big gap between researcher's knowledge and industry practices. An important goal of software development and software testing is to deliver quality software that meets or exceeds user expectations. Generally these two processes are independent resulting in some redundant and unnecessary effort. Integration of these two processes provide a better framework for coordination between the individual goals as well as for removal of redundancy between the common goals of the two processes.

Software development is a highly error prone process in which error may be injected at any of the generic phases through different sources. Detection and correction of errors early in the development life cycle is less costly than in later phase. Testing is one of the ways to detect an error, which on integrating within the development process may lead to identification of errors at each generic phase.

It is generally accepted that the only way to completely guarantee the correctness of a program is to execute it on all possible inputs, which is usually impossible or at least impractical. Thus, systematic testing techniques generate a representative set of test case to provide adequate testing for the program. Test cases are organized into test suites. Test suites are usually organized based on the kinds of test cases. Developing and organizing a test suite so that it is correct, observable and adequate is one of the greatest challenges in testing. There are two general forms of testing namely program-based and specification-based. Program-based testing is called White Box or glass-box testing, while specification-based testing is said to be Black Box. These techniques in program-based testing, certain aspects of code of a program, such as statements, branches, data dependencies, or paths are used to select test case. In specification based testing, test cases are designed to show that a program satisfies both functional and non-functional specification. This type of test cases are usually generated manually based on requirements derived form the specifications.

To validate program units, most systematic testing technique is used. In object-oriented programming, these units include object and classes, where classes are often seen as the smallest units of an object-oriented program that should be tested. Here, the unit testing in object-oriented programming means class testing, which is actually derived by the operations encapsulated by the classes and state behavior of the class. A range of program-based testing methods for testing classes have been proposed and used for testing object-oriented programs.

When unit tests are combined or integrated into subsystems, additional testing is required. Integration testing focuses on interfaces between units and subsystems. Integration testing of object-oriented programs can be very difficult due to the lack of hierarchical control structure, which makes it different to use the traditional top-down or bottom-up integration strategy. Instead, object-oriented integration testing can be performed by testing clusters of collaborating classes or components or by exploiting an underlying class dependency diagram to determine the test order of classes. Most integration testing methods are specification-based and program-based methods are also being developed.

Several testing models have been proposed in literature to test object-oriented software. Hetzel's Parallel Testing model as shown in figure 8.5(a) is one of them. Similar to the object-oriented software development model, the object-oriented software-testing model consists of the micro and macro processes. The macro process of testing is the controlling framework for the micro process of testing, and runs in parallel with the macro process of development. At the macro level, there are six stages of parallel testing of object-oriented software, described briefly as follows:

**Requirement Testing:** Requirement testing is performed during requirement gathering and analysis, and removes inadequacies if any. It involves the validation of methods and process to collect requirements, checking for the completion and consistency among the requirements. As requirements are laid out, test cases simulating common use cases or scenarios can be noted and referenced via those use cases. That is, there should exist traceability between the requirements and the test cases.

**Design Testing:** In order to ensure that the proposed architecture is correct, consistent, and complete, design should be tested. Testing at this phase will verify for the coverage of all requirements by the design.

**Unit Testing:** Testing at this phase will verify whether the individual piece or unit of object-oriented software system conform to their specifications and expected behavior.

**Subsystem Testing:** A set of logical related Units are tested at this phase of testing. All the verified individual piece of units is tested as a collection.

**Integrated System Testing:** In this phase of testing, the integrated system is tested for functionality, performance, reliability, and human factors.

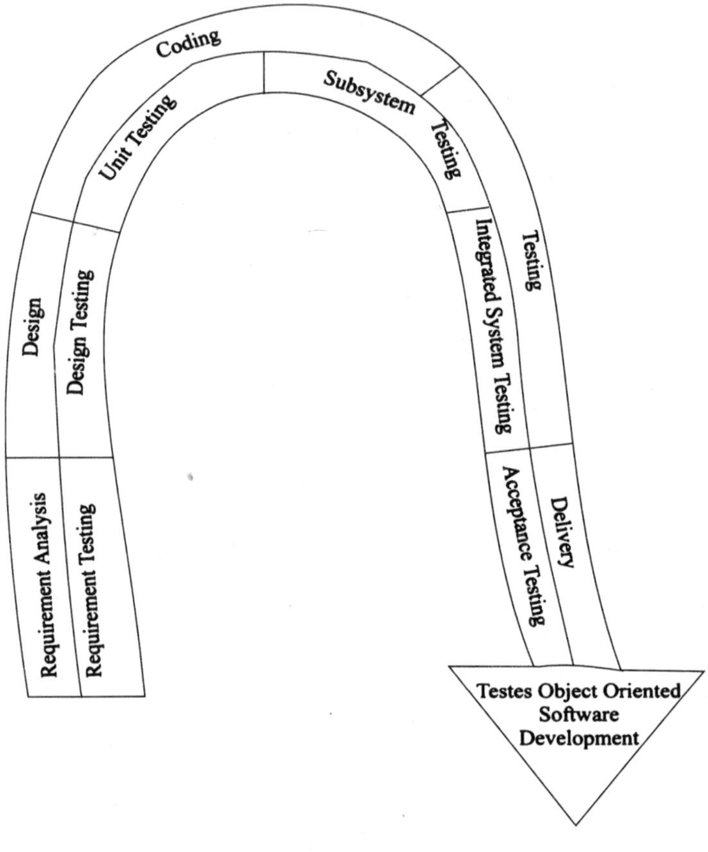


**Figure 8.5(a): Integrated Testing Life Cycle for Object-oriented Software**

**Acceptance Testing:** It is performed once the software is declared ready by the development team. This testing is done by an external organization.

The micro process of object-oriented software testing as shown in figure 8.5(b), controls an iterative and incremental micro process of object-oriented software development, and reflects daily activities. It is itself an iterative process including the following four major activities:

**Plan:** This phase will identify 'what needs to be tested during each iteration of testing'. All the necessary acquisitions will be specified and checked for their availability. Testability level will be predicted on the basis of the identified acquisitions, and phase wise testability will be estimated.

**Design:** This is one of the major activities during testing to facilitate the design of test cases to provide enough coverage.

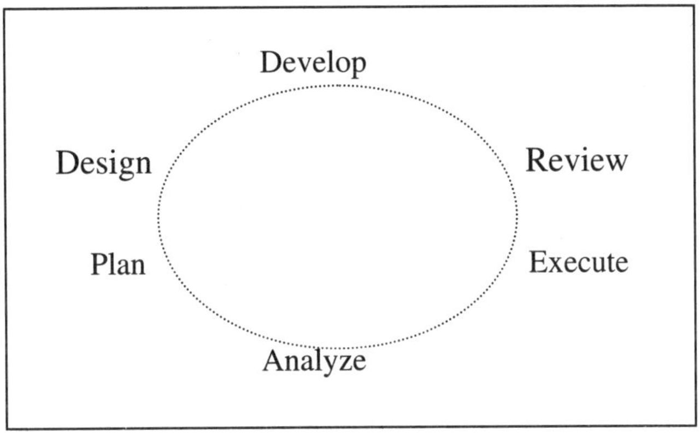


**Figure 8.5(b): Object-oriented Testing: Micro Process**

**Develop:** In this phase, each test sets defined, test cases are defined precisely.

**Review:** In this phase, developed test cases are reviewed for suitability, validity and reliability.

**Execute:** This is the final stage involved in execution of the defined test cases. The result of this stage facilitates with needed information on behavior discrepancies.

Analyze: This phase facilitates to analyze the execution behavior to uncover on faults.

## 8.6 OBJECT-ORIENTED SOFTWARE TEST STRATEGY

There does not appear any standard methodology for the testing of object-oriented software, which incorporates strategies for class hierarchy testing and polymorphism. Harrold,, McGregor and Fitzpatrick proposed the hierarchical testing strategy, which indicates the necessity of testing from a root class down the derivation tree to a leaf class. However, this strategy does not indicate what testing has to be performed for completeness.

In object-oriented program, a test strategy can be defined in the order of unit testing and integration testing of the classes. The test order problem for the classes in an OO program can be stated as finding an order to test the classes so that the effort required is minimum. Harrold did a pioneering work on object-oriented testing, and described a class testing methodology that utilizes the hierarchical nature of classes related by their inheritance relationships to reduce test overhead. Reusing the test information for a parent class and incrementally updating it to guide the testing of the subclasses accomplishes this. It is evident from various experiments that significant amount of efforts can be reduced with well designed hierarchies having large amount of functionality defined at the top level and modifications are done at lower level. The methodology includes the following activities:

- Designing a test suite for the base classes having no parents is chosen.
- Testing each member function and interactions among member functions.
- Defining a new subclass that inherits it's parent's testing history in addition to inheriting attributes.
- Incrementally updating the inherited testing history to reflect differences from parent, which leads to a testing history of subclass? A subclass's testing history indicates which test case must be run to test the subclass. Therefore it can be used as a guide for the execution of the test cases.
- Identifying new attributes in the subclass that must be tested along with inherited attributes that must be retested.
- Retesting the inherited attributes in the context of the subclass by identifying and testing their interactions with newly defined attributes in the subclass.
- Identifying test cases in the parent class's test suite that can be reused to validate the subclass and attributes of the subclass, which require new test cases.

Kung et al proposed another way of defining object-oriented software test strategy. Object-oriented testing begins by evaluating the correctness and consistency of the OOA and OOD models. Object-oriented testing strategy changes the concept, the 'unit' broadens due to encapsulation; integration focuses on classes and their execution across a 'thread' or in the context of a use case; and validation to be performed using conventional black box techniques.

There are several testing strategies that can be used during integration testing of object-oriented software. Generally the collaboration of objects is being tested by sending messages through subassembly testing. Scenarios and use cases have been widely used in system and software development to specify system requirements. However, scenarios can also be used for testing. Scenario testing, generally, tests the interaction of several objects that fulfill a specified use case.

Inheritance testing is very difficult part of the testing. It tests the polymorphism of a class hierarchy. While classification testing tests the relationship between instances of classes. Errors during inheritance and classification testing may be introduced because of the wrongly inherited class. The error may also be encountered if generic classes are improperly instantiated. Integration testing aims to find errors and inconsistencies in the integration of components. Errors with aggregation testing may occur if components are missing, and if there are inconsistent components.

For object-oriented software, unit testing is the process of testing a class and integration testing is the process of testing combinations of classes to discover interface errors between the classes. The traditional unit and integration levels of testing do not fit well in the case of object-oriented programs. Unit testing of subprograms cannot be mapped onto testing of individual object's routines: taken in isolation, the body of one routine typically consists of a few lines of code; its behavior is meaningless unless analyzed in relation to other routines and their joint effect on a shared state. Hence, any significant unit to be tested cannot be smaller than the instantiation of one class.

Unit test focuses on the verification of the smallest building blocks of software design - the class, ensuring that they work correctly in isolation, while integration testing is performed to uncover errors when two or more components are combined to execute their functionalities. While unit testing focuses on units in isolation, integration testing focuses on interactions and interrelationships among the units. Top-down, bottom-up, and sandwich testing are the conventional integration testing strategies. These testing strategies assume that there is a definable 'top' to the system. However, many object-oriented systems do not have a top in the sense of a main unit. This means that there is a need to devise new integration testing strategies.

Thread based testing integrates the set of classes required to respond to one input or event Use-case based testing integrates the set of classes required to respond to one use case and shows that every use case is executable within the system with and without user interface Testing a set of classes composing a system or subsystem is called interclass testing. Interclass testing usually performed during integration. Interclass testing applies to clusters of classes to be tested incrementally considering class interactions, polymorphic calls, and exception handling. Typically, such classes are not stand-alone entities, but mutually cooperate in several ways. These relationships among classes are a fundamental characteristic of object-oriented systems and define the nature of the interactions among objects at runtime. Different classifications exist of interclass relationships. Few of the phase wise testing strategies are discussed in the following section.

## 8.7 REQUIREMENT TESTING

Requirement testing is one of the neglected areas in the software development world. There is little evidence that software organizations do test the results of requirement collection and analysis at all. There is an scarcity of requirement testing tools. With the advent of modern software engineering techniques, testing requirements has become an essential element of software testing. It is still an ad hoc and informal process. With the emergence of object-oriented technology, the analysis and design phases are merging with each other. Analysis aims at precisely and clearly defining a problem for which design indents to find a solution. Therefore, the objectives and the focus of each phase are well separate. Figure 8.7 depicts the major activities in object-oriented software testing.

In object-oriented requirement analysis phase, analyst talks to the customer and delivers the information to the designers. Result of requirement analysis includes the description of the function of the system, along with the statement about performance and required resources. These results may be captured and produced through various formal and informal documents including the following described briefly.

**Prototyping:** This is one of the common elements of object-oriented analysis used to properly define the exact parameters of the problem. It helps customers to understand what they want in the desired system

**Graphical User Interface:** It is one of the elements of the delivered system through which customer interacts with the system. Through this documentation, different screen layouts are shown to the customers for their feedback and subsequent improvement.

**Requirement Specification Document:** This document includes all the requirements for desired system, and serves as the contractual agreement between customer and development team regarding the desired properties and behavior of expected system.

**Domain Object Model:** This model is used to capture the problem domain, in which object refers to a tangible thing for the customer of the desired system.

**Use Cases:** Use cases are the roadmap in which the desired system may be used. This is one of the most useful tools for capturing the exact behavior of the desired system.

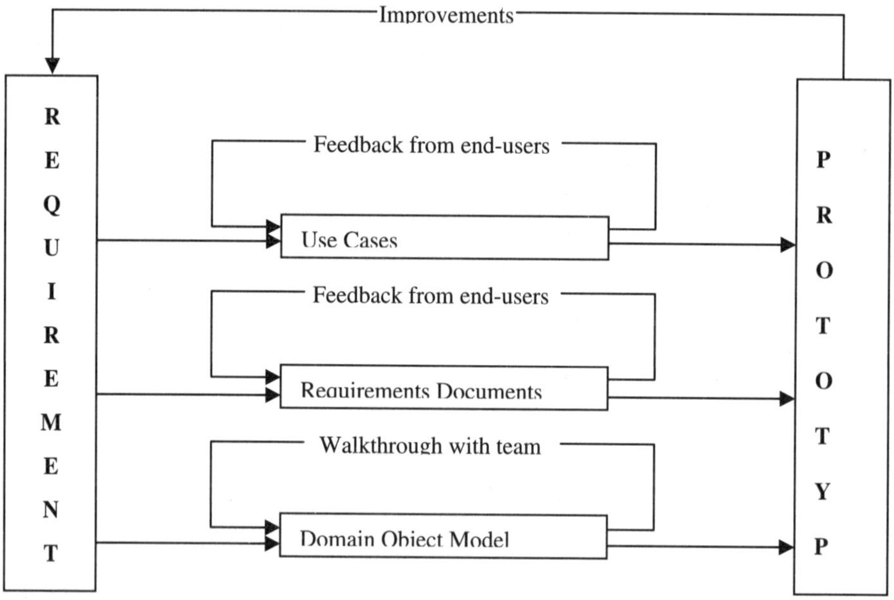


**Figure 8.7: Object-oriented Requirement Testing**

Major objective of the requirement testing is to verify and validate the results from the analysis phase in order to insure the quality of the requirements gathered. Testing at this phase helps in eradicating errors in the process early in the development life cycle. Testing requirements involves following issues:

**Correctness:** This is one of the most fundamental assumptions for designers. Since requirement analysis is the very first step in the software development process and makes the foundation for other activities in the process, the statements captured in various requirement analysis tools should be correct. Correctness of requirements eliminates the ambiguity in customer's true wishes. It ensures that the vague words in requirements statements are replaced.

**Completeness:** Requirements may be classified as functional, performance, resource and quality requirements. Functional requirement specifies the desired behavior of the system under specified input conditions. Performance requirements outline the response times of the system for various functionalities. Resource requirement illustrates the resources available for the desired system. Reliability and robustness of the system under specific conditions are specified by the quality requirements. Problems stated in all these kinds of the requirements should be completely and correctly specified.

**Consistency:** Consistency is another major aspect of requirement testing, and outlines the contradictory requirements. These incongruent requirements must be caught and eliminated during requirement testing. Redundant requirements specified by the users may be interpreted differently by different designers or implementers. Therefore, the redundant requirements should also be eliminated during requirement testing.

Requirement testing helps in detecting the inadvertent bugs in a software system at an early stage of software development life cycle. A little to no attention is given in practice, because of the lack of the software engineering standards and procedures, and insight into the benefits of requirements testing. Since, there are almost no formal tools available for collecting and analyzing requirements, it is generally captured in ambiguous English language.

The general approach for requirement testing is gathering feedback from the end-users through validating the use cases, the requirements specification documents, and showing the users the prototype. This feedback from the end-users addresses the major issues in requirement testing. Analysis guides the users in identifying the key elements of the requirements documents. A domain object model is created to capture the problem domain in a more formal manner. The feedback is then reviewed by the development team to fill up the holes in the requirement documents. Development team follows Walkthrough to improve the created domain model by eliminating incorrect relationships between classes and their stated behavior.

## 8.8 DESIGN TESTING

A design is a process that starts from a study of a domain problem leading to some formal documentation. At the first instance it may result in a model of the domain problem by formally capturing and representing the user's requirements and hence paving the way for a conceptual relation. It may serve well as a communication medium between the designer and the user on the one end, and act as a basis for implementation on the other end. This is evident that the software design is an important stage spanning the whole software lifecycle, not only for software development but also for re-developing legacy systems. It is concerned with accurately mapping the requirements from the analysis stage to logical models for implementation.

Testing software designs warrants significant effort, commitment, resources, and viable approach. Testing of design is not easy, however. There are no executables available till the prototypes are created. Testing of the transformation from analysis model to design model for completeness and correctness is mainly a manual process. Figure 8.8 depicts the major activities in object-oriented software testing.

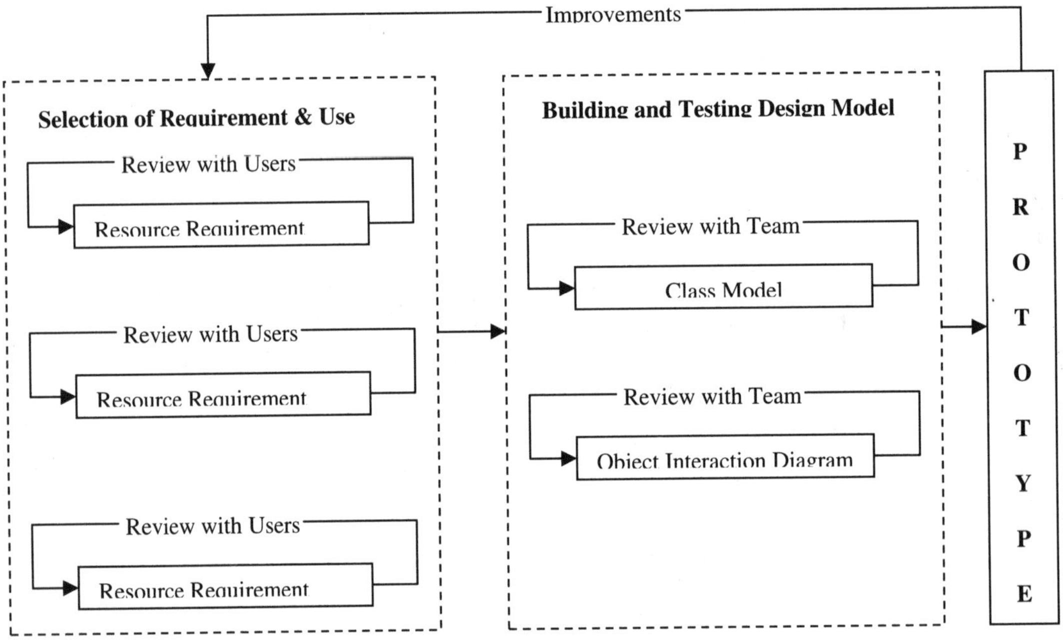


**Figure 8.8: Object-oriented Design Testing**

As object-oriented software development is maturing into an engineering discipline, more emphasis is being given to the testing of designs. An object-oriented design is not a strict top-down design of the functionalities of a system, rather it results in a set of collaborating objects working together to achieve a common goal. Focus of Design testing is normally to uncover the design faults of software, by looking mainly for features described as follows:.

**Completeness:** One of the important attributes of a good design is its complete solution to all the problems it is supposed to solve. A design solution should be able to meet the performance requirement, resource requirement as well as functional requirements. Therefore, another major objective of testing design is to ensure whether the design meets all its functional, resource, and performance requirements.

**Feasibility:** A design solution should be realizable for good design. Design testing ensures whether the design is possible to be implemented within the specified time and economic framework.

**Correctness:** A design should be verified for all its inputs and outputs. It is said to be correct if its input and output relation can be proved to be true or false.

Software complexity and required effort is the major issue during design testing. The lack of good approaches for testing object-oriented designs and scarcity of tools in the marketplace reveals a major challenge in design testing. Design testing may be conducted either manually or it may be automated. Manual techniques include design review, design walkthrough, and design inspection. Some of the formal verification techniques are based on inductive assertions, function semantics, and explicit semantics. Design walkthrough and prototyping are the two major approaches followed by the industry professionals in testing object-oriented software design. These are discussed in the following section.

**Design Walkthrough:** Design review to ensure that the designers did not miss any functional, performance, or resource requirements, is a powerful mechanism for identifying design errors. Several approaches are reported in literature that can be taken for design testing using Walkthrough. One of the best approaches is to find the best design through competition. In this approach, more than one design teams are involved in designing the system independently for the given time period. At the end, the designs are presented to a judging committee to find out the best design. Another possible approach for testing designs is to create a set of test cases for each iterations of the design. The review teams can Walkthrough the design using the test cases to determine whether a certain design property conforms to the output of the test case. Such models are used in testing database design configurations, transaction sequence, response time and user interfaces.

**Prototyping:** Another effective technique to validate the design is to transform the design into a physical shape which is called a prototype of that design. This approach gives designers the ability to choose between many designs alternatives. Prototyping also provides confidence in a designer's design decision, and supports designer's arguments about why a particular design decision was preferred. It gives designers the ability to physically see the behavior of a certain design property and discover erroneous design decisions at an early stage of software development life cycle.

## 8.9 UNIT TESTING

In object orientation, class is the basic building block for constructing software, and hence become the most natural unit of testing. Class-level testing requires substantial effort in the engineering of highly robust and reliable object-oriented software. Unit testing represents the most practical possible design, and provides the best form of documentation for classes. It determines when a class is done, and gives a developer confidence in the code. Unit testing is generally performed by the person who has developed the unit. The prerequisite for unit testing is the motivation to improve the quality of the unit. Unit testing is the part of software development and must be performed completely and honestly. It has been proved that adequate testing the methods of a class in isolation dose not ensure that a class has been adequately unit tested. A class should be tested with three different perspectives including functional, structural, and interaction.

Unit testing verifies the smallest building block of software design, and assures that individual parts of a complex system work correctly in isolation. Motivation and objectives for testing a class in isolation includes the following:

**Correctness:** Correctness is one of the basic objectives of class testing to furnish confidence that the intended logic has been correctly encapsulated in the class. Correctness of a class includes correctness of behavior and correctness of implementation. Correctness of behavior verifies whether the class delivers on the services that it promises. On the other hand, correctness of implementation can be determined by verifying whether the code conforms to its specifications, and objects deviate from its specified behavior.

**Completeness:** One of the major aspects of unit testing is to ascertain for necessary logic to be present in the class. Completeness may verify whether the class has all the necessary functions associated and available as the public interface of the class under test. It also verifies whether the method of a class completely execute its specified responsibility.

**Early Testing:** Unit testing usually tests a class prior to its integration with another class, which leads to early bug recovery, and reduces the cost of fixing bugs at the post design stages. It also provides confidence in the classes and objects during integration and system testing.

This is a well-understood fact that testing individual methods in a class cannot ensure that the class has also been adequately tested. There is a need to introduce some other testing strategy, which not only tests the interaction of methods in a class but also considers the state of an object during testing. Functional testing validates the black box view of methods in a class and exercises a member function as a standalone entity. The state of an object at a single instance of time is equivalent to the aggregate state of each of its data members at that instance. Methods of a class manipulate data members using various mechanisms. The correctness of a class verifies whether the data member represents the intended state of an object, and the member functions are correctly manipulating the representation of the object.

In general, there should be a unit test for every public method of a class. However, methods with very straightforward functionality do not require unit tests. A good guideline to follow is to write a unit test whenever there is a need to comment some behavior in the code, and put the unit tests in the same package as the associated classes being tested. This type of organization allows each unit test to call methods and reference variables that have access modifiers of package or protected in the class being tested. It is also recommended to avoid using domain objects that are specific to an application in unit tests.

## 8.10 INTEGRATION TESTING

Traditional approaches to integration testing include top-down, bottom-up, and sandwich approaches. Integration testing requires careful planning. Inheritance testing begins as soon as class is inherited from another. Integration strategies are needed at class level, component level, subsystem level, and system level. Classes require integration of methods. Various types of class interaction mechanism demand different testing strategies. System integration demands different types of integration testing strategies.

Integration testing is performed to catch any errors including internal and external errors, timing errors, and throughput errors, when two or more individually developed components are combined to execute their functionalities. In object-oriented system, the possible integrations may be (i) integration of members into a single class; (ii) integration of two or more classes through inheritance; (iii) integration of two or more classes through containment; (iv) integration of two or more classes to form a component; and (v) integration of components into a single application.

The primary objective of Integration testing is to ensure that individually developed units or components correctly interact to achieve the desired functionality of a system. In order to validate each component, the very first step is to ensure the availability of skeleton of the system as early as possible. As the skeleton of the system is available, individual parts of that skeleton are aggregated to ensure whether they interact correctly. Finally, it ensures whether all the parts interact reasonably well.

In traditional software development approaches, Integration testing focuses on the correct interfaces between multiple units. On the other hand, in object-oriented software behavior of the combined units has a state associated with it. Several integration testing strategies has been observed in literatures. Some of the pertinent once are as follows:

**Execution-Base Integration Testing:** It is used to reveal erroneous interaction of units by tracing their execution.

**Value-Based Integration Testing:** It is used to execute the interaction of units by employing certain values. Boundary value testing and equivalent class partitioning testing strategies fall under this category.

**Function-Based Integration Testing:** It is used to validate the functionality of components while they interact.

## 8.11 SYSTEM TESTING

Integrated system testing is a major phase of software testing life cycle. In addition to identifying errors at earlier stages, the distribution of testing also reduces the marathon effort required in a single stage. Integrated system testing includes sanity testing, documentation testing, performance testing, stress testing, and boundary testing.

System testing is a type of testing exercised during the last phase of software development life cycle. In the broader tem, system testing may be considered to include both the Integrated System testing and Acceptance testing. Depending upon the organization and software under test, alpha, beta, or any other form of testing may be complemented with this testing approach. One of the major objectives of system testing is to determine whether the software system is ready for its intended users by observing its behavior over a period of time. It tests the completely integrated system for conformance to requirements.

System testing must exercise all the functional, performance, stress, and resource requirements of a system. A functionally correct system may fail during its performance testing or stress testing. The major objective of system testing is to test all the aspects of software system completely and satisfactory. The basic strategy at this stage is to scrutinize the system under test for all possible combinations of data under all possible conditions, though it is infeasible in terms of time and resources. In order to resolve the dilemma, following verifications are required during system testing:

**Conformity Test:** One of the major tasks during system testing is to cover all requirements to ensure that all types of system requirements are exercised to ascertain their conformity.

**Data Coverage Test**: Another step during system testing is to judiciously select test data that cover a broad range of usages. It employs significant data values including valid data, invalid data, and boundary values to test the requirements in the first step.

For system testing, only abstract state of the objects needs to be defined. The abstract state of an object is domain-based, instead of value-based, and should be tested in unit and integration testing. It is not feasible to test state of an object during system testing.

## 8.12 CONCLUSION

Software testing is the center piece of software quality assurance. Irrespective of the lifecycle model used for software development, software has to be tested. Efficiency and quality are best served by testing software as early in the lifecycle as possible, with full regression testing whenever changes are made. The importance of software testing and its implications with respect to software quality cannot be overemphasized. Software testing is also responsible for bug prevention. The pragmatic goal of software testing is to discover bugs. Software testing also asserts to discover symptoms caused by bugs, and to provide clear diagnoses so that bugs can be easily corrected. When the software system is general buggy, automation does not show its advantage, and manual testing is effective. For manual testing, a good set of test outline is needed to assure the quality of software testing.

Object orientation gave a new direction to software development in terms of software reusability, extendibility, interoperability and reliability. Object-oriented software is developed in iterative and recursive increments. Three fundamental characteristics required for an object-oriented approach: Encapsulation, Polymorphism and Inheritance. Information hiding is a way of designing such that only the subset of the module's properties, its public interfaces, is known to users of the module. Polymorphism is an important concept that allows the building of a flexible system. Inheritance is a form of reuse that allows a programmer to define objects incrementally by reusing previously defined objects as the basis for new objects.

Unfortunately, the advantages of object-orientation become potential disadvantages when testing is considered. Object-oriented software is more difficult to test than the traditional structural programs and has to deal with new problems introduced by the powerful new features of object-oriented languages during testing. These features result in the so-called 'delocalized plan' in which several member functions from possibly several object classes are invoked to achieve an intended functionality. Ample number of work has been done to establish some basic software testing principles. Since object-oriented programs emphasize encapsulation of internal object state and interaction between objects through method calls, it is natural to base a testing strategy on scenarios of object usage. A test strategy, in object-oriented program, can be defined as the order to unit testing and integration testing of the classes. The test order problem for the classes in an object-oriented program can be stated as finding an order to test the classes so that the effort required is minimum. A very difficult part of object-oriented testing is inheritance testing, which tests the polymorphism of a class hierarchy. Scenario testing is used to test the interaction of several objects that fulfill a specified use case. For object-oriented software, unit testing is the process of testing a class, while integration testing is the process of testing combinations of classes to discover interface errors between the classes.

## KEY TERMS

Object-oriented Software, Reusability, Extendibility, Interoperability, Reliability, Traditional Procedural Language, Object-oriented Testing, Information Hiding, Class, Objects, Instance, Inheritance, Coupling, Sensitivity, Antiextensionality, Antidecomposition, Anticomposition, Errors, Faults, Failures, Frequent Testing, Testing Plan, Test Strategy, Test Suite, Subassembly Testing, Scenario Testing, Inheritance Testing, Classification Testing, Aggregation Testing, Unit Testing, Integration Testing, Thread Based Testing, Use-Case Based Testing, Interclass Testing

## REVIEW QUESTIONS

## OBJECTIVE TYPE QUESTIONS

1. Stronger coupling between artifacts leads to_________.
   (a) More difficult to understand individual artifacts
   (b) More easier to understand individual artifacts
   (c) Lager the sensitivity of changes and defect propagation
   (d) None of the above
2. ___________ means that all that is seen of an object is its interface.
   (a) Encapsulation
   (b) Coupling
   (c) Inheritance
   (d) Polymorphism

3. Functions are used in_________.
   (a) Procedural paradigm only
   (b) Object-oriented paradigm only
   (c) Both procedural and object-oriented paradigm
   (d) None of the above
4. In ________________ programmer can define the own data type.
   (a) Object-oriented paradigm
   (b) Procedural paradigm
   (c) Both procedural and object-oriented paradigm
   (e) None of the above

## SHORT ANSWER TYPE QUESTIONS

1. How can object-oriented software increase software reusability and extendibility? Discuss.
2. Software testing is the critical element to software quality. Comment.
3. Review the literatures on object-oriented testing. Write a paper emphasis on early estimation of testing efforts for object-oriented software.
4. Critically examine the difference between object-oriented paradigm and traditional paradigm, and prepare a comparative chart with reference to different parameters.
5. Propose a strategy to test object-oriented software. Take a very small module of object-oriented software; test it with reference to the proposed strategy.
6. Propose object-oriented testing metric suite to be used in early stage of development life cycle to estimate the testing effort.
7. How can you measure test adequacy for object-oriented software? Write down the test adequacy criteria, and justify it.
8. How can integration testing help in testing object-oriented software? Describe in detail.
9. There is a need to retest the classes in the context of overriding and overridden methods, even if the overriding and overridden methods are semantically similar. Justify why?

## DESCRIPTIVE QUESTIONS

1. How does object-oriented software differ from traditional software?
2. How can software testing prevent bugs in software?

3. What do you mean by software reusability? How can it assist in testing object-oriented software?
4. List the misconceptions regarding the testing.
5. 'Procedural program is task-based'. Justify.
6. Write down the test adequacy axioms.
7. List the objectives of testing object-oriented software. How to achieve these objectives?
8. How can you design tests that systematically uncover different classes of errors with the minimum amount of time and effort?
9. Why are testing and maintenance of object-oriented software costly?
10. Distinguish between complete testing and enough testing.
11. What is the purpose of performing frequent testing in case of testing object-oriented software?

## REFERENCES


1. J. Voas. PIE: A dynamic failure-based technique. IEEE Transactions on Software Engineering, 18(8):717–727, August 1992.
2. J. Voas and K. Miller. Semantic metrics for software testability. Journal of Systems and Software, 20:207–216, March 1993.
3. R.Wheeldon and S. Counsell. Power law distributions in class relationships. In Proceedings of the Third International Workshop on Source Code Analysis and Manipulation. IEEE Computer Society, September 2003.
4. Huo Yan Chen, T.H. Tse, and T.Y. Chen, "TACCLE: a methodology for objected-oriented software testing at the class and cluster levels", ACM Transactions on Software Engineering and Methodologv, Vol. 10, No. 1, 2001, pp.56-109.
5. Huo Yan Chen, T.H. Tse, F.T. Chan, and T.Y. Chen, "In black and white: an integrated approach to class level testing of object-oriented programs", ACM Transactions on Software Engineering and Methodology, Vol. 7, No. 3, 1998, pp.250-295.
6. Watson and T. McCabe. Structured testing: A software testing methodology using the cyclomatic complexity metric. T. NIST Special Publication 500-235, National Institute of Standards and Technology, Washington, D.C., 1996.
7. Huo Yan Chen, T.H. Tse, and Y.T. Deng, "ROCS: an object-oriented class-level testing system based on the Relevant Observable Contexts technique", Information and Software Technologv, Vol. 42, No. 10,2000, pp. 677-686. Programming, Osbome McGraw-Hill, Berkeley, Califomia, 1992.

8. M. J. Harrold and J. D. McGregor, \Toward a testing methodology for object-oriented software systems," Department of Computer Science, Clemson University.

9. M. J. Harrold, J. D. McGregor, and K. J. Fitzpatrick, \Incremental testing of object-oriented class structure", Proc. of 14th International Conf. on Software Engineering, pp. 68 - 80, 1992.

10. M.D. Smith, and D.J. Robson, "A framework for testing object-oriented programs", Journa/ of Object-Orientedprogramming, Vol. 5, No. 3, 1992, pp. 45-53.

11. D. Kung, J. Gao, P. Hsia, Y. Toyoshima, and C. Chen, \A test strategy for object-oriented systems," Proc. of Computer Software and Applications Conference, pp. 239 { 244, Dallas Texas, August 9{11, IEEE Computer Society, 1995.

12. D. Kung, J. Gao, P. Hsia, Y. Toyoshima, C. Chen, Y.S. Kim, and Y. Song, \Developing an object-oriented software testing and maintenance environment", Communications of the ACM, Vol. 38, No. 10, pp. 75 { 87, October 1995.

13. D. Kung, Y. Lu, N. Venugopalan, P. Hsia, Y. Toyoshima, C. Chen, J. Gao, \Object state testing and fault analysis for reliable software systems," Proc. Of 7th International Symposium on Software Reliability Engineering, White Plains, New York, Oct. 30 - Nov. 2, 1996.

14. V. Aho, J. E. Hopcroft and J. D. Ullman, \Data Structures and Algorithms," Addison-Wesley Publ. Comp., 1983.

15. R. Doong and P. Frankl, \The ASTOOT approach to testing object-oriented programs," ACM Transactions on Software Engineering and Methodology, Vol. 3, No. 2, pp. 101 - 130, April 1994.

16. M. D. Fraser, K. Kumar, and V. Vaishnavi, \Strategies for incorporating formal speci_cations," CACM Vol. 37, No. 10, pp. 74 { 86, 1994.

17. D. Ho_man and P. Strooper, \ClassBench: a framework for automated class testing," Software Practice and Experience, Vol. 27, No. 5, pp. 573- 597, 1997.

18. P. Hsia, X. Li, C. Hsu, D. Kung, \A Technique for Selective Revalidation of OO Software," Journal of Software Maintenance, to appear.

19. P. Jorgensen and C. Erickson, \Object-oriented integration testing," CACM Vol. 37, No. 9, pp. 30 - 38, September 1994.

20. H. Kim and C. Wu, \A class testing technique based on data bindings," Proceedings of the 1996 Asia Pacific Software Engineering Conference, pp. 104-109, Soul, South Korea, December 1996.

21. T. Korson and J. D. McGregor. \Understanding object- oriented: a unifying paradigm." CACM Vol. 33, No. 9, pp. 40 - 60, Sept. 1990.

22. D. Kung, J. Gao, P. Hsia, J. Lin and Y. Toyoshima, \Design Recovery for Software Testing of Object-Oriented Programs," Proc. of the Working Conference on Reverse Engineering, pp. 202 - 211, Baltimore Maryland, May 21 - 23, IEEE Computer Society Press, 1993.

23. L. C. Briand, S. Morasca, and V. Basili. An operational process for goal-driven definition of measures. IEEE Transactions on Software Engineering, 28(12):1106–1125, December 2002.

24. S. Chidamber and C. Kemerer. A metrics suite for object-oriented design. IEEE Transactions on Software Engineering, 20(6):476–493, June 1994.

25. Van Deursen and T. Kuipers. Building documentation generators. In Proceedings of the International Conference on Software Maintenance (ICSM'99), pages 40–49. IEEE Computer Society, 1999.

26. R. Freedman. Testability of software components. IEEE Transactions on Software Engineering, 17(6):553–564, June 1991.

27. Weyuker, E.J. , Axiomatizing software test data adequacy, IEEE TSE, Vol. SE-12,1986

28. Weyuker, E.J. ,The evaluation of program-based software test data adequacy criteria, Communications of the ACM, Vol.31,1988

29. David S. Rossenblum and Elaine J.Weyuker, Using Coverage Information to Predict the Cost-Effectiveness of Regression Testing Strategies, IEEE Tran. On Software Engineering, Vol.23, No.3., 146-156, March 1997

30. Daniel Hoffman, ClassBench:a Framework for Automated Class Testing, Software – Practice and Experience, Vol.27(5),573-597, May, 1997.

31. Shyam R. Chidamber, and Chris F. Kemerer, A Metric Suite for Object-oriented Design, IEEE Tran. On Software Engineering, Vol.20, No.6., 476-493, June 1994

32. The complexity of object modules in OO technology, C.F. Mai, Computer Science, p19-23,Vol.23,1996

33. D. Kung, N. Suchak, P. Hsia, Y. Toyoshima, and C. Chen, \On object state testing," Proc. Of COMPSAC'94, pp. 222 { 227, IEEE Computer Society Press, 1994.

34. D. Kung, J. Gao, P. Hsia, F. Wen, Y. Toyoshima, and C. Chen, \Change impact identification in object-oriented software maintenance," Proc. of IEEE International Conference on Software Maintenance, pp. 202 - 211, 1994.

35. D.L. Parnas, \A technique for software module specification with examples," CACM May, 1972.

36. Parrish, Allen S., Richard B. Borie and David W. Cordes, \Automated Flow Graph-Based Testing of Object-oriented Software Modules," Journal of Systems Software, no. 23, 1993.

37. D. E. Perry and G. E. Kaiser, \Adequate testing and object-oriented programming," Journal of Object-Oriented Programming, Vol. 2, pp. 13 - 19, January-February 1990.

38. R. Poston, \Automated testing from object models," CACM Vol. 37, No. 9, pp. 48 - 58, September 1994.

39. G. Rothermel and M. J. Harrold, \Selecting Regression Tests for Object-oriented Software,", Proc. Of IEEE International Conference On Software Maintenance, pp. 14-25, 1994.

40. M. D. Smith and D. J. Robson, \Object-oriented programming - the problems of validation," Proc. IEEE Conference on Software Maintenance, pp. 272 - 281, 1990.

41. [23] E. Soloway et al, \Designing documentation to compensate for delocalized plans," CACM Vol. 31, No. 11, pp. 1259 - 1267, Nov. 1988.

42. C. D. Turner and D. J. Robson, \The state-based testing of object-oriented programs," Proc. Of IEEE Conference on Software Maintenance, pp. 302 -310, 1993.

43. E. J. Weyuker, \Axiomatizing software test data adequacy," IEEE Trans. on Software Eng., Vol. SE-12, No. 12, pp. 1128 - 1138, 1986.

44. E. J., Weyuker, \The evaluation of program-based software test data adequacy criteria," CACM, Vol. 31, No. 6, pp. 668 - 675, 1988.

45. N. Wilde and R. Huitt, \Maintenance support for object-oriented programs," IEEE Trans. on Software Eng., Vol. 18, No. 12, pp. 1038 - 1044, December 1992.

46. Boris Beizer, Software Testing Techniques, Second Edition, International Thomson Computer Press,1990

47. Secrets of Software Quality, Craig Kaplan, Ralph C.,Victor Tang, McGraw-Hill, Inc. 1994

48. Daniel Tkach, Richard Puttick, Object Technology in Application Development, Second Edition, Addision-Wesley Publish Company,1996

49. Osten O.S. & Robert L.G. An ISO 9000 Approach to Building Quality Software, Prentice Hall PTR, New Jersey, 1996

50. Mayers, G.J., The Art of Software Testing, John Wiley &Sons Inc. New York, 1979, Chinese version, by Zhiying Zhou, Renjie Zheng, Tsing Hua Press£¬1985

51. Roger S. Pressman, Software Engineering--A Practitioner's Approach, 2nd edition, McGraw-Hill . 1992,

52. Bill Hetzel, The Complete Guide to Software Testing, Second Edition, John Wiley & Sons, Inc.,1988

53. Robert B.G., Practical Software Metrics for Project Management and Process Improvement, PTR Prentice Hall, New Jersey,1989

54. Practical Software Engineering 2nd edition, Ren-jie Zheng,, Tsing Hua Press 1997

55. Software test process model--POCERM, Liu Chao, Jin Maozhong,, Jr. of BUAA, p.56-p.60,Vol.23 No.1, Feb.1997

56. Jin Ling-zi, Process in Testing Object-Oriented Software, Computer Research & Development, Jan., 1985

57. R. Binder. Design for testability in object-oriented systems. Comm. of the ACM, 37(9):87–101, 1994.
58. R. Binder. Testing Object-Oriented Systems: Models, Patterns, and Tools. Addison-Wesley, 2000.
59. B.W. Boehm. Software Engineering Economics. Prentice Hall, Englewood Cliffs, NJ, 1981.
60. L. C. Briand, J.W. Daly, and J. K.W¨ust. A unified framework for coupling measurement in object-oriented systems. IEEE Tr. on Software Engineering, 25(1):91–121, 1999.
61. B. Henderson-Sellers. Object-Oriented Metrics. Prentice Hall, New Jersey, 1996.
62. ISO. International standard ISO/IEC 9126. information technology: Software product evaluation: Quality characteristics and guidelines for their use, 1991.
63. S. Jungmayr. Identifying test-critical dependencies. In Proceedings of the International Conference on Software Maintenance, pages 404–413. IEEE Computer Society, October 2002.
64. J. McGregor and S. Srinivas. A measure of testing effort. In Proceedings of the Conference on Object-Oriented Technologies, pages 129–142. USENIX Association, June 1996.
65. L.H. Putnam. A general empirical solution to the macrosoftware sizing and estimating problem. IEEE Tr. on Software Engineering, 4(4):345–61, 1978.
66. Rountev, A. Milanova, and B. G. Ryder. Fragment class analysis for testing of polymorphism in java software. In Proceedings of the 25th International Conference on Software Engineering (ICSE'03). IEEE Computer Society, 2003.
67. S. Siegel and N. J. Castellan Jr. Nonparametric statistics for the behavioral sciences. McGraw-Hill Book Company, New York, 1988.
68. Imran Bashir & Amrit L Goel, Testing Object-oriented Software-Life Cycle Solutions, Springer Verlag 1999.

## USEFUL LINKS

http://www.csis.gvsu.edu/~erickson/OO_Testing/ACMpaper.ps.Z
http://www.mozilla.org/
http://www.dbai.tuwien.ac.at/ftp/papers/overbeck-diss.ps.Z
http://jakarta.apache.org/cactus/index.html.

## Chapter 9

# Software Fault Tolerance

*Program testing can be used to show the presence of bugs, but never to show their absence.*
**Edsger Dijkstra**

## OBJECTIVES

A successful completion of this chapter will enable the reader for the following:

- To understand and describe fault tolerance in general.
- To appreciate the need, importance and significance of software fault tolerance.
- To define and understand the software failure.
- To realize the facts related to various checks available to detect error.
- To describe techniques of software fault tolerance.
- To visualize the different perspectives of detecting software errors.
- To understand the importance of software system having high degree of availability.
- To select a test process to determine the fault tolerant of a system.
- To analyze the co-relation of software fault tolerance with dependability of software system.
- To understand dependability chain.
- To realize the need of early identification of problems in software development life cycle..
- To understand the protection means for software failure.
- To differentiate software faults and common design faults.
- To define error-based testing techniques.
- To understand fault seeding mechanism used in fault-based testing.
- To appreciate the need, importance and significance of mutation testing.
- To apply fault tolerance concepts to test fault tolerance of system.

## 9.1 INTRODUCTION

Software systems are being used for applications that require high reliability, tolerance and dependability. They are used in areas where failures could be catastrophic, as for controlling nuclear plants, air traffic, space exploration and defense. Software can also have small unnoticeable errors or drifts that may culminate into a disaster. In 1991, on $25^{th}$ February, during the Gulf War, the chopping error that missed 0.000000095 second in every $10^{th}$ of a second, accumulating for 100 hours, made the Patriot missile fail to intercept a Scud missile. As a result twenty-eight lives were lost. In the same year, after changing just three lines of code in a signaling problem containing more than a million lines of code, the local telephone systems in California along with the Eastern seaboard came to a standstill. This shows that fixing problems may not necessarily make the software more reliable and qualitatively better.

Software security relates entirely and completely to quality. It appears to be more important to think about security, reliability, availability, and dependability from the beginning. Although researchers are well aware of the software security problem, they have just focused on late life-cycle testing stuff. Early identification of a problem is required for assuring effective eradication and minimizing the impact in the prospective phases. In general, software problems include bugs in the code and architectural inadequacies in the design. In the past decade researchers have paid too much attention to bugs and very little to architectural flaws, perhaps mainly due to the possibility of bugs being visible during the program execution.

Software systems have to possess a high degree of availability in many cases. The desired requirement can be ensured only if software systems have been made fault tolerant and have been tested thoroughly. Testing is a costly process and, therefore, efficient techniques are needed for performing testing. Though testing cannot rule out the absence of faults, it should give some idea about the reliability of the system. The methodology used for testing should also help the tester in designing better test suites. Moreover it will be better if testing process is able to determine 'how fault tolerant a system is?'.

Software fault tolerance is strongly co-related with dependability of the software. In the chain of dependability threats, a fault activates an error and error propagates to become a failure. Dependability is a term that covers a number of useful requirements including Availability, Reliability, Safety, and Maintainability. These are described briefly as follows.

**Availability:** Availability of a software system is ability that the system is ready to be used immediately. In general, it refers to the probability that the system is operating correctly at any given moment and is available to perform its functions on behalf of its users. A highly available system is one that will most likely be working at a given instant in time.

**Reliability:** In contrast to availability, software reliability is defined in terms of a time interval instead of an instant in time. Reliability refers to the property that a system can run continuously without failure. A highly reliable system is one that will most likely continue to work without interruption during a relatively long period of time.

**Safety:** Safety refers to the situation that when a software system temporarily fails to operate correctly, nothing catastrophic happens.

**Maintainability:** Software maintainability refers to how easily a failed system can be repaired. A highly maintainable system may also show a high degree of availability, especially if failures can be detected and repaired automatically.

## 9.2 NEED FOR SOFTWARE FAULT TOLERANCE

Complex safety critical systems currently being designed and built are often difficult multi-disciplinary undertakings. Part of these systems is often a computer controlled subsystem. Services in today's communications networks are highly robust. The computer industry has historically been concerned about hardware failure. Today, however, with amazingly high hardware reliability, the emphasis is shifting to software. The increasing complexity of software causes growing concern about the robustness of system software components. Even when exceptional caution is taken, simple logic errors can result in catastrophic failures. Thus, protection from software failure becomes the central challenge. It is practically impossible to prevent software failure completely, and design approaches that virtually eliminate software failure can more than triple development costs and still not guarantee that failure will not occur.

In order to ensure that these systems perform as specified, even under extreme conditions, it is important to have a fault tolerant computing system; both hardware and software. To achieve hardware fault tolerance, it is necessary to replicate all critical hardware components. Not surprisingly, the same principle applies for Software Fault Tolerance. In order to protect a software application from software faults, the application's logic and data must be replicated. However, software must be tested hard to tolerate the faults of different type before deployment. This may facilitate to function normally even in adverse situations.

## 9.3 SOFTWARE FAILURE

To understand and develop the methods for software fault tolerance, it is required to understand what faults to tolerate. A series of failures in hardware components, subsystems or systems have been better tracked over the years and there is a larger body of experience in failure modes and effects analysis. However, the counterpart in software is far less understood. There is also need to have some clarity on software failure while discussing software fault tolerance. Different classes of software failures have been identified including the following:

**Transient:** Transient failure occurs with certain specific input to the software.

**Permanent:** Every time when the code is executed, the failure occurred is called permanent error.

**Recoverable:** If the failure in the system can be recovered without intervention, the failure is called recoverable failure.

**Unrecoverable:** If, while recovering a system after failure has detected, an operator intervention is needed, is called unrecoverable failure.

**Corrupting:** If the failure corrupts the system state or data, it is said to be corrupting failure.

**Non-corrupting:** If the error remains confined, the failure is called non-corrupting.

## 9.4 PRINCIPLES OF SOFTWARE FAULT TOLERANCE

The objective of software fault tolerance is to include safety features in the software design or source code to ensure that the software will respond correctly to input data errors and prevent output and control errors. Following section describes the phases of fault tolerance. These phases, taken together, provide the general means by which fault tolerance can be implemented to prevent faults which lead to system failure.

**Detect and correct the error:** Generally, software faults are because of the design faults occurred while a designer either misunderstood a specification or simply made a mistake. Since in the modern system, complexity is pushed into the software portion, software fault has become a very common phenomenon. It is evident from the literature that 60-90 percent of current computer errors are because of the software faults. Software faults may also occur in hardware. But these faults can be masked using a combination of current software and hardware fault tolerance techniques. In order to tolerate a fault in a system its effects must be detected first. Faults cannot be directly detected by a system. The manifestation of the fault will generate errors somewhere in the system. Thus, it will be worthwhile to check for error first to detect the erroneous state during fault tolerance of software. Several mechanisms are available for error detection, few are described briefly as follows:

**Replication Checks:** Replication checks make use of matching components with error detection based on comparison of their outputs.

**Timing Checks:** Timing checks are applicable to systems and modules whose specifications include timing constraints, including deadlines. The checks can be developed to look for deviation from the acceptable module behavior on the basis of these constraints. A type of timing check that can be used to monitor for satisfactory behavior, is called watchdog timer. Watchdog timers can also be used to detect lost components.

**Reversal Checks:** In this type of the check, the output of a module is used to compute the corresponding inputs based on the function of the module. If the computed inputs do not match with the actual input, an error is detected. This type of the checking is usually applied to modules whose inverse computation is relative straightforward.

**Coding Checks:** Coding checks use redundancy in the representation of information with fixed relationships between the actual and the redundant information. Error detection is based on checking those relationships before and after operations.

**Reasonableness Checks:** Reasonableness check detects errors by using known semantic properties. These properties can be based on the requirements or the particular design of a module.

**Structural Checks:** Structural check uses known properties of the data structure. Structural checks could be made more effective by augmenting data structures with redundant structural data. Structured methodology may be used to develop error detection strategies. Error detection strategy can also be developed in an ad-hoc fashion by experienced designers. Fault trees have been proposed as a design aid, to identify general classes of failures and conditions that can trigger those failures, in the development of fault detection strategies.

**Confinement and Assessment of Damages:** Whenever an error occurred, much more of the system state may be suspect than that initially discovered to be erroneous. It is necessary to make an attempt to assess the extent to which the system state has been damaged, while dealing with the detected errors. This is because of the likely delay between the manifestation of a fault and the detection of its erroneous consequences. Since the assessment will depend on run-time measures and mechanisms used, the system designer concerning damage confinement should take the proper decisions in this regard.

**Fault Treatment and Continuous Service:** The system will return to an error free state after detecting and recovering the system. When the system is in error-free state, techniques should be used to enable the system to continue, so that it can provide the desired services. The detection of error does not serve the purpose, and may not identify faults. Therefore, fault treatment will attempt to locate the remaining faults accurately, and reconfigure the rest of the system to avoid the fault.

## 9.5 TECHNIQUES OF SOFTWARE FAULT TOLERANCE

Software is generally an integral component of all computing based systems. Indeed, the major proportion of the complexity of most systems is to be found in the software. It is somewhat surprising that the provision of tolerance to software faults has rarely been considered necessary. One reason for this may have been the absence of suitable fault tolerance techniques. Two main methods have been proposed for structuring a software system and providing software fault tolerance: recovery blocks, and N-version programming. They should be regarded as complementary to, and not as a substitute for, other techniques (such as testing), which remove faults before the software is put into operation.

**Recovery Block Scheme:** The Recovery Block technique combines the basics of the checkpoint. This simple method, developed by Randell, is a multi-version software fault technique. Multi-version fault tolerance is based on the use of two or more versions of a piece of software, executed either in sequence or in parallel. During recovery block technique, different version of a software component is tried after an error is detected. Before executing a version of a piece of software, checkpoints needs to be created, because it will help to recover the state after a version fails to provide a valid operational starting point for the next version if an error is detected. The recovery block scheme provides a unifying framework for the implementation of fault tolerant software, incorporating strategies for error detection by the acceptance test and interface checks provided by the underlying system. As yet there is not a great deal of experience with the use of recovery blocks for providing software fault tolerance.

**N-Version Programming:** Software system, where each module is made up to N-different implementation is called N-version software system. N-Version programming is a multi-version technique in which all the versions are designed to satisfy the same basic requirements. Each variant accomplishes the same task, but hopefully in a different way. Each version then submits its answer to voter or decider which determines the correct answer and returns that as the result of the module. The decision of output correctness is based on the comparison of all the outputs. Several generic decision algorithms are available to select the correct output in N-version programming technique. This characteristic makes it different from Recovery Block approach, which requires an application dependent acceptance test. Table 9.5(a) shows the difference between recovery method and N-version programming technique. N-version software system may also be used to overcome the design faults present in most software by relying upon the design diversity concept. N-version can only be successful and successfully tolerate faults if the required design diversity is met.

**Table 9.5(a): Recovery Blocks approach Vs N-version method**

| Recovery Block approach | N-version method |
|---|---|
| • It has been extended to include concurrent execution of the various alternatives. | • It has always been designed to be implemented using N-way hardware concurrently. |
| • It requires that each module build a specific adjudicator. | • A single decider may be used. |

**N Self-Checking programming:** A multiple software versions are combined with structural variations of the Recovery Blocks and N-Version Programming. This combined technique is called N Self-Checking programming technique. In this case, the versions and the acceptance tests are developed independently from common requirements. N Self-Checking model uses the separate acceptance tests for each version, while Recovery Block executes and tests the versions sequentially or in parallel but the output is taken from the highest-ranking version that passes its acceptance test. It is not a rigorously described method in the literature, but rather a more ad hoc method used in some important system. Self-checking software has been implemented in some extremely reliable and safety-critical systems already deployed in our society. Self-checking software are the extra checks, often including some amount check pointing and rollback recovery methods added into fault tolerant systems. Sequential execution requires the use of checkpoints, and parallel execution requires the use of input and state consistency algorithms. This model also uses an application independent decision to select a correct output similar to N-version programming.

**Exception Handling:** Exception handling is a single-version fault tolerance, which is based on the use of redundancy applied to a single version of a piece of software to detect and recover from faults. Exception handling is a process of the interruption of normal operation to handle abnormal responses. As for as the software fault tolerance is concerned, exceptions are signaled by the implemented error detection mechanisms as a request for initiation of an appropriate recovery. Several exception triggering events for a software component have been listed in the literature, including: interface exceptions, internal local exceptions, and failure exceptions defined in table 9.5(b).

**Table 9.5(b): Exception Triggering Events**

| Interface Exceptions | Local Exceptions | Failure Exceptions |
|---|---|---|
| • It is signaled by a component when it detects an invalid service request.<br>• It is triggered by the self-protection mechanisms of a module.<br>• It is meant to be handled by the module that requested the service. | • It is signaled by a module when its error detection mechanisms find an error in its own internal operations.<br>• It is handled by the module's fault tolerant capabilities. | • It is signaled by a module after it has detected an error which its fault processing mechanisms have been unable to handle successfully. |

## 9.6 FAULT-BASED TESTING METHODS

This type of testing allows for designing test cases on the client specification or code. It tries to identify possible faults in the area of design or code that may lead to errors. For each of these faults a test case is developed to 'flush' the errors out. These tests also force each line of codes to be executed. Several fault-based method have been identified in the literature including the followings:

**Error-Based Testing:** Error based testing defines classes of errors as well as inputs that will reveal any error of a particular class, if it exists.

**Fault Seeding:** Fault seeding implies the injection of faults into software prior to testing. Based on the number of these artificial faults discovered during testing, inferences are made on the number of remaining 'real' faults. For this to be valid the seeded faults must be assumed similar to the real faults.

**Mutation Testing:** Mutation testing injects faults into code to determine optional test inputs.

**Fault Injection:** Fault injection evaluates the impact of changing the code or state of an executing program or behavior of the software.

## 9.7 CONCLUSION

Software security relates entirely and completely to software quality. It appears to be more important to think about security, reliability, availability, and dependability at the beginning in the design, architecture, test and coding phases. The increasing complexity of software causes growing concern about the robustness of system software components. Even simple logic errors can result in catastrophic failures. Hence, protection from software failure becomes the central challenge.

Majority of system defects are shown to be software defects. It is practically impossible to prevent software failure completely, and design approaches that virtually eliminate software failure can more than triple development costs and still not guarantee that failure will not occur. In order to ensure that the system performs as specified, even under extreme conditions, it is important to have a fault tolerant computing system. Software fault tolerance is strongly related to dependability of the software, and is an immature area of research. One of the objectives of software fault tolerance is to include safety features in the software design or source code to ensure that the software will respond correctly to input data errors and prevent output and control errors. Software faults may be because of the design faults occurred while a designer either misunderstood a specification or simply made a mistake. Fault treatment will attempt to locate the undetected faults accurately, and will reconfigure the rest of the system to avoid the fault.

## KEY TERMS

Availability, Reliability, Safety, Maintainability, Dependability, Software Faults, Software Fault Tolerance, Software Failures, Replication Check, Timing Check, Reversal Check, Reasonableness Check, Structural Check, Recovery Block, N Version Programming, N-Self Checking Programming, Exception Handling, Fault Based Testing Methods, Error Based Testing, Fault Seeding, Mutation Testing, Fault Injection

## REVIEW QUESTIONS

## OBJECTIVE TYPE QUESTIONS

1. _________ propagates to become failure.
   (a) Fault
   (b) Error
   (c) Bugs
   (d) None of the above
2. Abilities that the system is ready to be used immediately is called______.
   (a) Reliability
   (b) Availability

(c) Dependability

(d) Safety

3. ___________ is defined in terms of time interval instead of an instance of time.

(a) Availability

(b) Reliability

(c) Safety

(d) Maintainability

4. In situation that when a software system temporarily fails to operate correctly, nothing catastrophic happens______.

(a) Safety

(b) Reliability

(c) Maintainability

(d) None of the above

5. Faults can be masked using______.

(a) Software fault tolerance techniques

(b) Hardware fault tolerance techniques

(c) Combination of both software and hardware fault tolerance techniques.

(d) None of the above

6. A check developed to look for deviation from the acceptable module behavior on the basis of constraints is______.

(e) Replication check

(f) Timing check

(g) Reversal check

(h) Loading check

## SHORT ANSWER TYPE QUESTIONS

1. Define dependability.
2. How does software security relate to software quality?
3. What do you mean by architectural flaws?
4. Distinguish between availability and dependability.
5. How can testing facilitate for assuring reliability of the system?
6. What do you mean by fault tolerance?
7. How can testing determine system's is fault tolerance level?

8. How can software fault tolerance be related to dependability of the software?
9. How can you estimate the safety of the software system?
10. Write the need, importance and significance of the software fault tolerance.
11. Distinguish between recoverable and unrecoverable errors.
12. Elaborate the principles of software fault tolerance.
13. Can a system directly detect fault? Why or why not?
14. How can coding checks differ from structural checks?
15. How can structural methodology be used to develop error detection strategies?
16. What are fault trees? What is the purpose of introducing fault trees?
17. How cn you facilitate fault treatment, in software systems?
18. Write the technical availability for software fault tolerance.
19. Distinguish between recovery block and N-version programming.
20. How can recovery block schcmc work in software fault tolerance?
21. Distinguish between fault based and error based testing methodologies.
22. What do you mean by fault seeding? How does it differ from mutation testing?
23. How can fault injections help in software fault tolerance?

## DESCRIPTIVE QUESTIONS

1. Testing is costly process and therefore efficient techniques are needed for performing testing. Justify.
2. Correlate software fault tolerance with software dependability.
3. Establish a relationship amongst fault, error and failure.
4. Propose a metric that can be used to quantify safety of software system.
5. Why are highly maintainable software system show a high degree of availability? Justify.
6. Do an exhaustive review on software fault tolerance. Write a paper proposing a model for object oriented software fault tolerance.

## REFERENCES


1. Ying shi, 'Fault Tolerance Computing – Draft', Carnegie Mellon University, 18-849b, Dependable embedded systems, 1999.
2. Offutt and S. Lee, "An Empirical Evaluation of Weak Mutation," IEEE Transactions on Software Engineering, vol. 20, no. 5, May 1994.

3. E. Delamaro, C. Maldonado and A.P. Mathur, "Interface Mutation: An Approach for Integration Testing," IEEE Transactions on Software Engineering, Vol. 27, No. 3, March 2001, pp. 228-247.
4. A .J. Offutt and J. Pan, "Automatically Detecting Equivalent Mutants and Infeasible Paths," The Journal of Software Testing, Verification, and Reliability, Vol 7, No. 3, September 1997, pp. 165-192.
5. N. Storey, 'Safety critical computer systems', Harlow, England, Addison Wesley, 1996.
6. J. Gray and D. P. Siewiorek, 'High availability computer system', IEEE computer, 24(9):39-38,1991.
7. 91 P. Nesi and T. Querci. Effort estimation and prediction of object-oriented systems. The Journal of Systems and SOBware, Vo1.42, pp.89-102, 1998.
8. J. Rumbaugh, M. Blaha, W. Premerlani, F. Eddy, and W. Lorensen. Object-Oriented Modeling and Design. Prentice Hall International, Englewood Cliffs, New Jersey, 1991.
9. H. M. Sneed. Estimating the costs of software maintenance tool. In Proc. International Conference on Software Maintenance, PP. 168- 18 1, Opio, France, October 17-20 1995.
10. M. R.Lyu, 'Software Fault Tolerance', ed., chichester, England, John Wiley and sons, Inc., 1995.
11. D.E. Eckardt and L.D. Lee, "A Theoretical Basis for the Analysis of Multiversion Software Subject to Coincident Errors," IEEE Transaction on Software Engineering, vol, SE-11, no. 12, December 1985, pp. 1511-1517.
12. B. Littlewood, P. Popov and L. Strigini, "Assessing the Reliability of Diverse Fault-Tolerant Software-based Systems," Safety Science, vol40, 2002, pp.781-796.
13. K. Kim, M. A. Vouk and D.F. McAllister, "An Empirical Evaluation of Maximum Likelihood Voting in High Inter- Version Failure Correlation Conditions," Proceedings of 7th International Symposium on Software Reliability Engineering, October 1996, pp 330-339.
14. R. Zhao, Research on Software Testing Methodologies, Ph.D. thesis, Chinese Academy of Science, 2001.
15. Chris Inacio, 'Software fault tolerance', Carnegie Mellon University, 18-849b, Dependable embedded systems, 1998.
16. J. Devale and P. Koopman, 'An N-version Approach to Measuring operating system Robustness', in FTCS-29, 1999.
17. J.C. Munson, T.M. Khoshgoftaar. The Detection of Faultphone Programs. IEEE Transactions on Sofhvare Engineering ~01.18, n.5, pp.423-432, 1992.

18. D. Thomas and 1. Jacobson. Managing object-oriented software engineering. In Tutorial Note, TOOLS'89, International Conference on Technology of Object-Oriented Languages and Systems, pp.52, Paris, France, 13-15 Nov. 1989.
19. S. W. Y. Lee, B. Liang and F. Wang. Measuring the coupling and cohesion of an object oriented program based on information flow. In Proc. International Conference on Sofware Quality, Maribor; Slovenia, 1995.
20. H. Zuse. A Framework of Software Measurement. Walter de Cruyter, Berlin, New-York, 1998.
21. Ian & John Pugh, 'Carrier-grade, high availability computing platforms for voice and data networks', FTCS proceedings, 1999.
22. Gautam Kumar Saha, 'Beyond the conventional techniques of Software Fault Tolerance', Ubiquity, vol.4, issue 47, Jan 28- feb3, 2004.
23. Jiantao Pan, 'A Conceptual Framework for System Fault Tolerance', http://www.stlabs.com
24. T. Anderson & P. A. Lee, 'Fault Tolerance: Principles and Practice' PHI 1981.
25. James Hamilton, 'Fault Avoidance Vs. Fault Tolerance: Testing Doesn't Scale', Microsoft SQL Server development, HPTS 1999 position paper.
26. D.R. Kuhn and M.J. Reilly, "An Investigation of the Applicability of Design of Experiments to Software Testing," Proc. 27th NASA/IEEE Software Eng. Workshop, Dec. 2002.
27. S.R. Dalal, A. Jain, N. Karunanithi, J.M. Leaton, C.M. Lott, G.C. Patton, and B.M. Horowitz, "Model-Based Testing in Practice," Proc. Int'l Conf. Software Eng., 1999.
28. B. Smith, M.S. Feather, and N. Muscettola, "Challenges and Methods in Testing the Remote Agent Planner," Proc. Fifth Int'l Conf. Artificial Intelligence Planning Systems, 2000.
29. J. Pan, "The Dimensionality of Failures—A Fault Model for Characterizing Software Robustness," Proc. Int'l Symp. Fault-Tolerant Computing, June 1999.
30. M. Rebaudengo, M. Sonza Reorda, M. Vilante, 'A new approach to software implemented fault tolerance', IEEE, Latin American Test workshop, 2002.
31. 'Fault Tolerance', IEEE Trans. Software July/ August 2001.
32. Simon Dyck and Dacid Sloane, 'Software Testing', A work from software Eng. Network research center from university of Calgary, Canada, 1999.
33. Boris Beizer, 'Software Testing Techniques', 2nd ed., Van Nostrand, Reinhold, New York, 1990.
34. R. Brownlie, J. Prowse, and M.S. Phadke, "Robust Testing of AT&T PMX/ StarMail Using OATS," AT&T Technical J., vol. 71, no. 3, pp. 41-47, May/ June 1992.
35. D.M. Cohen, S.R. Dalal, J. Parelius, and G.C. Patton, "The Combinatorial Approach to Automatic Test Generation," IEEE Software,, vol. 13, no. 5, pp. 83-88, Sept. 1996.

36. V.N. Nair, D.A. James, W.K. Erlich, and J. Zevallos, "A Statistical Assessment of Some Software Testing Strategies and Application of Experimental Design Techniques," Statistica Sinica, vol. 8, no. 1, pp. 165- 184, 1998.

37. D.R. Wallace and D.R. Kuhn, "Failure Modes in Medical Device Software: An Analysis of 15 Years of Recall Data," Int'l J. Reliability, Quality and Safety Eng., vol. 8, no. 4, 2001.

38. M. Jones, "What Really Happened on Mars Pathfinder Rover," RISKS Digest, vol. 19, no. 49, Dec. 1997.

39. J.L. Lions, "Ariane 5, Flight 501, Report of the Inquiry Board," European Space Agency, July 1996.

40. G. Slabodkin, "Software Glitches Leave Navy Smart Ship Dead in the Water," Government Computer News, July 1998.

41. B. Stevens, L. Moura, and E. Mendelsohn, "Lower Bounds for Transversal Covers," Design, Codes, and Cryptography, vol. 15, pp. 279-299, 1998.

42. The CRC Handbook of Combinatorial Designs. C.J. Colbourn and J.H. Dinitz, eds., CRC Press, 1996.

43. A.W. Williams and R.L. Probert, "A Measure for Component Interaction Test Coverage," Proc. ACS/IEEE Int'l Conf. Computer Systems and Applications (AICCSA 2001), pp. 304-311, June 2001.

44. C. Cheng, A. Dumitrescu, and P. Schroeder, "Generating Small Test Suites for Non-Uniform Instances ," Third Int'l Conf. Quality Software, Nov. 2003.

45. P.E. Ammann, P.E. Black, and W. Majurski, "Using Model Checking to Generate Tests from Specifications ," Proc. Second IEEE Int'l Conf. Formal Eng. Methods (ICFEM '98), pp. 46-54, Dec. 1998.

46. L. Briand and D. Pfahl, "Using simulation for assessing the real impact of test coverage on defect coverage," Proceedings of 10th IEEE International Symposium on. Software Reliability Engineering, Nov. 1999, pp. 124-157.

47. M.Chen, M.R. Lyu, and E. Wong, "Effect of Code Coverage on Software Reliability Measurement," IEEE Transactions on Reliability, vol. 50, no. 2, June 2001, pp. 165-170.

48. Y.K. Malaiya, N. Li, J. M. Bieman, and R. Karcich, "Software Reliability Growth with Test Coverage," IEEE Transactions on Reliability, vol. 51, no. 4, December 2002, pp. 420-426.

49. A.J. Offutt, A. Lee, G. Rothermel, R.H. Untch, and C. Zapf, "An Experimental Determination of Sufficient Mutant Operators," ACM Transactions on Software Engineering Methodology, vol. 5, no. 2, 1996, pp. 99-118.

50. M.R. Blackburn, R.D. Busser, A.M. Nauman, and R. Chandramouli, "Model Based Approach to Security Test Automation," Proc. Quality Week, June 2001.
51. K.C. Tai and Y. Lie, "A Test Generation Strategy for Pairwise Testing ," IEEE Trans. Software Eng., vol. 28, no. 1, pp. 109-111, 2002.
52. A.W. Williams and R.L. Probert, "Formulation of the Interaction Test Coverage Problem as an Integer Program ," Proc. TestCom Conf., pp. 283- 298, 2002.
53. M.B. Cohen, C.J. Colbourn, P.B. Gibbons, and W.B. Mugridge, "Constructing Test Suites for Interaction Testing," Proc. 25th Int'l Conf. Software Eng. (ICSE 2003), pp. 38-48, May 2003.
54. D.M. Cohen, S.R. Dalal, M.L. Fredman, and G.C. Patton, "The AETG System: An Approach to Testing Based on Combinatorial Design," IEEE Trans. Software Eng., vol. 23, no. 7, pp. 437-444, July 1997.
55. M. Lyu (ed.), Handbook of Software Reliability Engineering, McGraw-Hill, New York, 1996.
56. S. Rapps, E. J. Weyuker, "Selecting Software Test Data Using Data Flow Information," IEEE Transactions on Software Engineering, vol. SE-11, No.4, April 1985.
57. W. E. Howden, "Weak Mutation Testing and Completeness of Test Sets," IEEE Transactions on Software Engineering, vol. SE8, no. 4, July 1982, pp. 371-379.
58. P.G. Frankl, E.J. Weyuker, "An Applicable Family of Data Flow Testing Criteria," IEEE Transactions on Software Engineering, vol. SE-14, No. 10, October 1988.
59. E.J.Weyuker, "The Cost of Data Flow Testing: An Empirical Study," IEEE Transactions on Software Engineering, v.16 n.2, February 1990, pp.121-128.
60. J.R. Horgan, S. London, and M.R. Lyu, "Achieving Software Quality with Testing Coverage Measures," IEEE Computer, vol. 27, no. 9, September 1994, pp. 60-69.
61. A.J. Offutt, G. Rothermel, and C. Zapf, "An Experimental Evaluation of Selective Mutation," Proceedings of the 15$^{th}$ International Software Engineering Conference, May 1993, pp. 1-107.
62. W.E. Wong and A.P. Mathur, "Reducing the Cost of Mutation Testing: An Empirical Study," The Journal of Systems and Software, vol. 31, no. 3, December 1995, pp. 185-196.
63. B. Randell and J. Xu, "The Evolution of the Recovery Block Concept," Software Fault Tolerance, M. R. Lyu (ed.), Wiley, Chichester, 1995, pp. 1-21.
64. A.A. Avizienis, "The Methodology of N-Version Programming," Software Fault Tolerance, M. R. Lyu (ed.), Wiley, Chichester, 1995, pp. 23-46.
65. J.C. Laprie, J. Arlat, C. Beounes, and K. Kanoun, "Architectural Issues in Software Fault Tolerance," Software Fault Tolerance, M. R. Lyu (ed.), Wiley, Chichester, 1995, pp. 47-80.

66. K.E. Grosspietsch, "Optimizing the Reliability of the Component-based N-version Approaches," Proceedings of International Parallel and Distributed Processing Symposium (IPDPS 2002), Fort Lauderdale, Florida, April 2002, pp. 138-145.

67. K.E. Grosspietsch, A. Romanovsky, "An Evolutionary and Adaptive Approach for N-version Programming," Proceedings of 27th Euromicro Conference, Warsaw, Poland, September 2001, pp. 182-189.

68. M. R. Lyu and Y. He, "Improving the N-version programming process through the evolution of a design paradigm," IEEE Transactions on Reliability, 42(2), June 1993, pp. 179-189.

69. B. Littlewood and D.Miller, "Conceptual Modeling of Coincident Failures in Multiversion Software," IEEE Transactions on Software Engineering, vol. 15, no. 12, December 1989, pp. 1596-1614

70. X. Teng and H. Pham, "A Software-Reliability Growth Model for N-Version Programming Systems," IEEE Transactions on Reliability, vol. 51, issue 3, September 2002, pp. 311-321.

71. F. Belli and P. Jedrzejowicz, "Fault-Tolerant Programs and Their Reliability," IEEE Transactions on Reliability, vol. 29(2), 1990, pp. 184-192.

72. B. Littlewood, P. Popov, and L. Strigini, "Modelling Software Design Diversity - a Review," ACM Computing Surveys, Vol. 33, No. 2, June 2001, pp. 177-208.

73. B. Littlewood, P. Popov, P. and L. Strigini, "Design Diversity: an Update from Research on Reliability Modelling," Proceedings of Safety-Critical Systems Symposium 21, Bristol, U.K., Springer, 2001.

74. L.J. White and E.I. Cohen, "A Domain Testing Strategy," IEEE Transactions on Sofiware Engineering", Vol. SE-6, No. 3, May 1980, pp. 247-257.

75. D.E.Eckhardt, Caglavan, Knight, Lee, McAllister, Vouk, Kelly, "An experimental evaluation of software redundancy as a strategy for improving reliability," IEEE Transaction of Software Engineering, vol 18, July 1991, pp 692-702.

76. M.R. Lyu, J.R. Horgan, S. London, "A Coverage Analysis Tool for the Effectiveness of Software Testing," Proceedings of ISSRE'93, Denver, November 1993, pp. 25-34.

77. R. Chillarege, I.S. Bhandari, J.K. Chaar, M.J. Halliday, D.S. Moebus, B.K. Ray, and M.Y. Wong, "Orthogonal Defect Classification – A Concept for In-Process Measurements," IEEE Transactions on Software Engineering, vol. 18, no. 19, November 1992, pp.943-956.

78. Y. K. Malaiya, L. Naixin, J. Bieman, R. Karcich and B. Skibbe, "The Relationship Between Test Coverage and Reliability," Proceedings of 5th International Symposium on Software Reliability Engineering, November 1994, pp. 186 -195.

79. M. Hitz, B. Montanzeri, Measuring coupling and Cohesion in Object-Oriented Systems, In Proc. Of the International Symposium on Applied Corporate Computing, Monterrey, Mexico. http://www.pri.univie.ac.at/ hitzlpaperslESEC95.p~

80. D.W. Hosmer, S. Lemeshow. Applied Logistic Regression. John Wiley & Sons, 1989.

81. H. F. Kaiser. The varimax criterion for analytic rotation in factor analysis. Psychometrika, vo1.32, pp.187-198, 1958.

82. W. Li and S. Henry. Object-oriented metrics that predict maintainability. The Journal of Systems Software, vo1.23,

## USEFUL LINKS

http://www.hpl.hp.com/techreports/98/HPL-98-06.html
http://portal.research.bell-labs.com/orgs/ssr/people/lyu/
http://www.ece.cmu.edu/~jdevale/
http://www.ece.cmu.edu/~koopman/des_s99/fault_tolerant/index.html
http://www.ece.cmu.edu/~koopman/des_s99/ultra_dependability/index.html

## Chapter 10

# Design Phase Mutation

*The greatest test of courage on earth is to bear defeat without losing heart.*
**Robert Green Ingersoll**

## OBJECTIVES

A successful completion of this chapter will enable the reader for the following:

- To understand mutation and mutation analysis.
- To realize the need and importance of mutation testing.
- To distinguish between live and killed mutants.
- To define equivalent mutants.
- To find out mutation score.
- To elaborate test data adequacy.
- To realize the importance of test data adequacy in optimizing the testing process.
- To understand fault-based technique.
- To analyze the efficiency of test data suite.
- To create mutant program for the segment of the program.
- To understand object oriented mutation in general.
- To elaborate object oriented mutation in design phase.
- To identify and define the mutant operators for object oriented software.
- To realize the importance and role of mutant operators in estimating software test adequacy.
- To analyze different object oriented mutant operators in early stage of development.
- To classify object oriented mutation operators.
- To define the process of mutation.
- To automate mutant generation mechanism.
- To write mutation operator of the given high-level diagram.
- To check the efficiency of generated mutants.

## 10.1 INTRODUCTION

Testing, as an integral process, provides a vital support during software engineering, but developing tests often take significant resources. Software testing involves exercising a program on a set of test case input values and comparing the actual output result with expected ones. One of the major tasks in testing is to select the minimal set of test cases sufficiently effective for revealing potential faults in a program. But the fundamental question arises how? The test strategies are faced with the problematic question like this, since exhaustive testing is usually not feasible in real life software application. It is evident that the evaluation criterion for test strategies may be used to measure the effectiveness of generated test cases. Mutation analysis is one of the evaluation techniques that assesses the quality of test cases by examining whether they can reveal certain types of faults. It is sometimes helpful to view software development as a 'root-finding' procedure where a designer submits an initial 'guess', and is iteratively tested for validity until a correct solution is converged upon. Initially, this principle was called 'meta-induction' and was used at a conceptual model for a new approach to software testing, called 'mutation analysis'.

Mutation testing is considered one of the promising testing techniques. It consists of creating a sample of faulty programs from the original program under test, by injecting faults. The set of similar programs, called 'mutants', is obtained after applying a single mutation operator to a single location in program. These mutants are run with an input data from a given test set. If a test set can distinguish a mutant form the original program, i.e. it produces different execution result, the mutant is said to be '*killed*'. Otherwise, the mutant is called a 'live' mutant. A mutant remains live because it is equivalent to the original program, that is, it is functionally identical to the original program although syntactically different, or the test set is inadequate to kill the mutant. If the mutant is 'equivalent mutant', it would always produce the same output, hence it can't be killed. Typically the equivalent mutants are distinguished approximately after testing or identified by manually. If a test set is inadequate, it can be improved by adding test cases to kill the live mutant. A test set which can kill all non-equivalent mutants, is said to be 'adequate' and the 'mutation score'. The adequacy of a test set, is measured as a ratio of the number of mutants killed over the total number of non-equivalent mutants, as follows:

$$\text{Adequacy of Test Set} = \frac{\text{Number of killed mutants}}{\text{Number of non-equivalent mutants}}$$

This can be used to examine the adequacy of test data for object-oriented software, and also to assist to optimize the testing process. This technique may be used to define a test set from the randomly generated test pool. It may be used in deciding whether to stop testing a product or to enhance the test set by constructing additional test cases. It is evident from the literature that the testing activities should be shifted to the possibly early parts of the development process. There is a processing need to enforce the mutation approach early in the development life cycle. Object-oriented mutation for testing the object-oriented program may be used to investigate the results of object-oriented mutation operators applied to the design level. Moreover, mutation operator needs to be evaluated in terms of their effectiveness of detecting faults in object-oriented software.

## 10.2 STRUCTURAL MUTATION

Mutation testing is a powerful technique for unit-testing software. However, it is quite computationally expensive, complicated and time consuming. It is a fault-based technique that aims particularly at measuring the effectiveness of test cases. Some times it is advisable to view the software development process as a roadmap where designer's initial guess is iteratively tested for validity through mutation analysis. It also aims to determine whether a software system has been adequately tested. In the process of root finding, the objective of testing may be to verify following:

- If the initial program is incorrect, the modified program can be obtained with the help of an operator. Fixing one or more faults in initial program may do this.
- If initial program is correct, this can be delivered as a final program.

Let's consider a program segment that finds the rank correlation to the first time where maximum values appear in the array, as follows:

```
x=1;
for (i=2; i<=3; i++)
{
        if (a[i] > a[x]
        {
                x=i;
        }
}
```

Now, some mutants may be produced by introducing simple modifications of the initial program, as follows:

**Mut-I**

```
x=1;
for (i=1; i<=3; i++)
{
        if (a[i] > a[x]
        {
                x=i;
        }
}
```

**Mut-II**

```
x=1;
for (i=1; i<=3; i++)
{
        if (i > a[x]
        {
                x=i;
        }
}
```

**Mut-III**

```
x=1;
for (i=2; i<=3; i++)
{
        if (a[i] >= a[x]
        {
                x=i;
        }
}
```

**Mut-IV**

```
x=1;
for (i=2; i<=3; i++)
{
        if (a[x] > a[x]
        {
                x=i;
        }
}
```

Further, consider a test-data suite for the segment as per the following table 10.2(a):

**Table 10.2(a): Test Data Suit**

| | a[1] | a[2] | a[3] |
|---|---|---|---|
| DT1 | 2 | 3 | 4 |
| DT2 | 2 | 3 | 2 |
| DT3 | 4 | 2 | 3 |

Table 10.2(b) shows the result after applying the test data to mutants: Mut-I, II, III and IV.

**Table 10.2(b): Killed and Live Mutants**

| | **P_Initial** | **P_Modified** | | | | **Killed Mutants** |
|---|---|---|---|---|---|---|
| | | **Mut-I** | **Mut-II** | **Mut-III** | **Mut-IV** | |
| **DT1** | 3 | 3 | 3 | 3 | 1 | **Mut-IV** |
| **DT2** | 2 | 2 | 3 | 2 | 1 | **Mut-II & IV** |
| **DT3** | 1 | 1 | 1 | 1 | 1 | **None** |

Table 10.2(c) shows the adequacy of test data suits used to test the program.

**Table 10.2(c): Adequacy of Test Data Suits**

| **Test Data Suit** | **Testing Adequacy** |
|---|---|
| DT1 | 25% |
| DT2 | 50% |
| DT3 | 100% |

By analyzing the efficiency of test data proposed, it is noticed that Mut-I and Mut-III are not killed by the test data selection. It may be deduced that the test is incomplete, and there is a need to add a new test data DT4= {2, 2, 3}, initial program will have x=1, and Mut –III will have 2, then this test data will kill Mut-III. While in case of Mut-I, Mut-I starts with i=1 and initial program starts with i=2, which has no impact on the result r. Mut-I is a secondary mutant which can't be distinguished from initial program by any test data. It is evident from the foregoing discussion that there is a need for mutant generation, test data selection and execution of all mutants in mutation technique.

## 10.3 OBJECT-ORIENTED MUTATION

Some research has been done on object-oriented mutation, development of mutation operators for object-oriented software and design specific operators for Java program. Mutation analysis is a fault-based technique with the aim of measuring the mutant of test data. It involves constructing a set of mutant programs, each of which is a copy of initial program under test that differs by exactly one mutation. A mutation may be defined as the single, syntactically correct change that is made by applying a predefined transformation rule, called a mutation operator, to the original program. Within this paper through object-oriented mutation will be understood applying such object oriented mutation operators which concern only those features of the program that directly refer to the construction and control of class and methods. Object-oriented features are fully represented in the object-oriented specification, for example, on HLD/LLD diagram. Therefore, our aim is to investigate the object-oriented operators, which can be applied to HLD/LLD of object-oriented software. We recognize the following groups of object-oriented mutation operators for the HLD/LLD.

**INH Operator (Inheritance Operator):** This operator changes the base class in the inheritance or omits the inheritance relation.

**ENC Operator (Encapsulation Operator):** This operator generates a mutant by changing the structure of class either by adding, and omitting or replacing the keywords (private, public, protected).

**COP Operator (Coupling Operator):** This mutant operator changes association between classes. It generates mutants by omitting the association, changing its type, navigation and replacing with another one pointing to the base or inherited class.

**COH Operator (Cohesion Operator):** This operator changes the functional strength of a class by changing the methods of class.

All the mutant operators described, attempt to alter the original program semantics through a simple syntactic changes. They target possibly flaws related to object-oriented concepts. These operators are supported to mimic common mistakes made by developers. They are chosen considering HLD/LLD prepared by medium experienced developers. These mutant operators have been applied on the HLD/LLD of 'Company' (Fig. 10.3). The following tables show some of the mutants of each class and the faults those mutants are intended to represent.

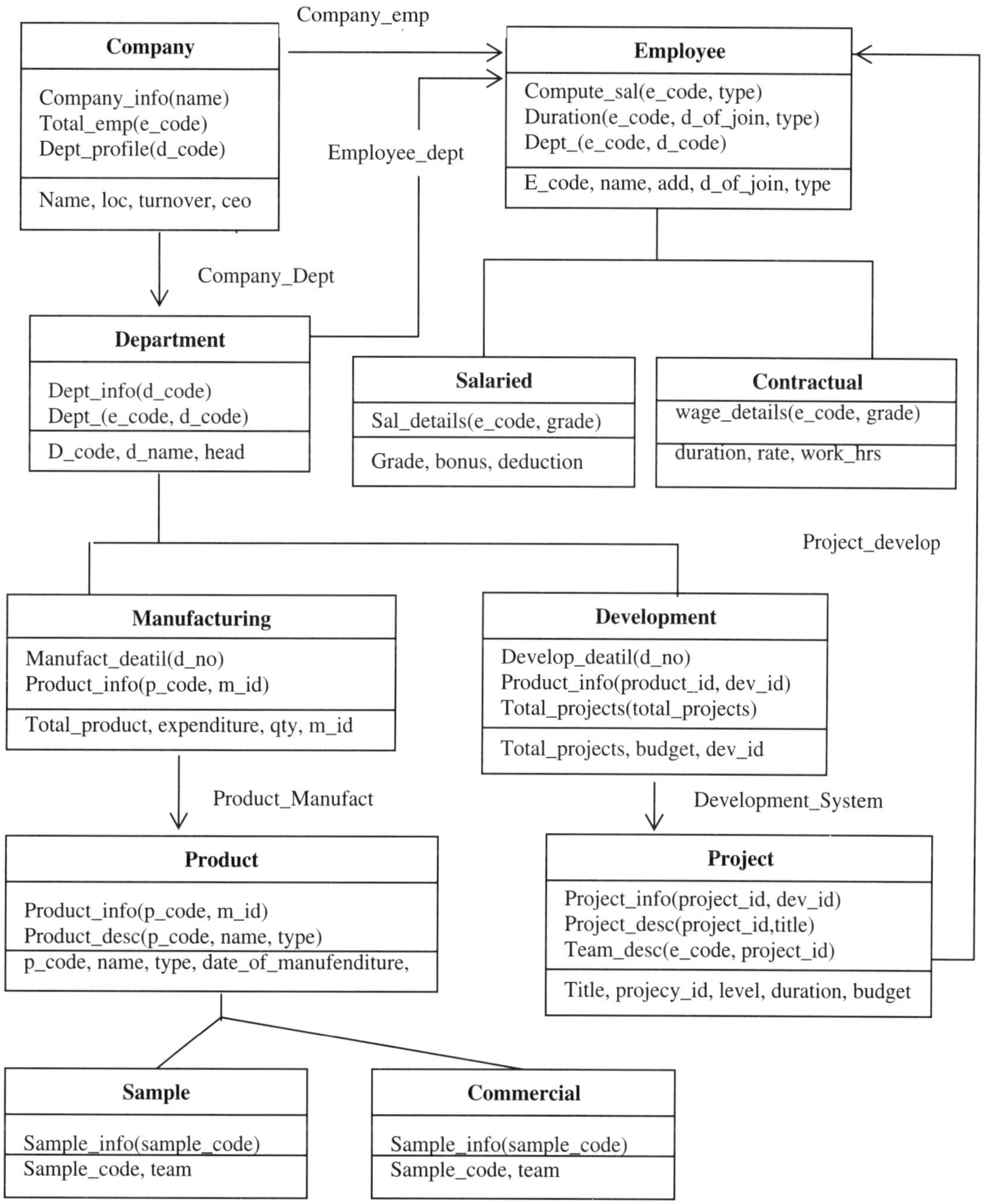


**Figure10.3: Class Diagram**

**Table 10.3(a): Mutants of Company (CM Stands for Company Mutant)**

| Mutants | Fault Description |
|---|---|
| CM1 (COP) | The association comp-emp directs from class employee to company |
| CM2 (COP) | The association comp_dept directs from class department to company. |
| CM3 (COH) | The method dept_profile is called instead of company_info |
| CM4 (COH) | The method company_info is called instead of dept_profile |
| CM5 (COH) | The method company_info is called instead of total_emp |
| CM6 (COH) | The method total_emp is called instead of company_info |
| CM7 (COH) | The method total_emp is called instead of dept_profile |
| CM8 (COH) | The method dept_profile is called instead of total_emp |

**Table 10.3(b): Mutants of Employee (EM Stands for Employee Mutant)**

| Mutants | Fault Description |
|---|---|
| EM1 (ENC) | Dept_() is protected in class employee |
| EM2 (COP) | The association comp_emp directs from employee to company |
| EM3 (COP) | The association emp_dept directs from employee to department |
| EM4 (COH) | Method compute_sal() is called instead of duration() |
| EM5 (COH) | Method duration()is called instead of compute_sal() |
| EM6 (COH) | Method duration()is called instead of dept_() |
| EM7 (COH) | Method dept_() is called instead of duration() |
| EM8 (COH) | Method dept_() is called instead of compute_sal() |
| EM9 (COH) | Method compute_sal() is called instead of dept_() |
| EM10 (COP) | The association project_develop directs from project to employee |

**Table 10.3(c): Mutants of Department (DM Stands for Department Mutant)**

| Mutants | Fault Description |
|---|---|
| DM1 (ENC) | Dept_info() is protected in calss department |
| DM2 (COH) | Method dept_info() is called instead of dept_() |
| DM3 (COH) | Method dept_() is called instead of dept_info() |
| DM4 (COP) | Association comp_dept directs from department to company |
| DM5 (COP) | Association emp_dept directs from employee to department |

**Table 10.3(d): Mutants of Manufacturing (MM Stands for Manufacturing Mutant)**

| Mutants | Fault Description |
|---|---|
| MM1 (COP) | The association product_manufact directs from product to manufacturing |
| MM2 (COH) | Method manufact_detail() is called instead of product_info() |
| MM3(COH) | Method product_info()is called instead of manufact_detail() |
| MM4 (INH) | Class manufacturing has no department as the base class |

**Table 8 Mutants of Development (DvM Stands for Development Mutant)**

| Mutants | Fault Description |
|---|---|
| DvM1 (COP) | The association development_system directs form project to development |
| DvM2 (COH) | Method develop_detail() is called instead of product_info() |
| DvM3 (COH) | Method product_info() is called instead of develop_detail() |
| DvM4 (COH) | Method develop_detail() is called instead of total_projects() |
| DvM5 (COH) | Method total_projects() is called instead of develop_detail() |
| DvM6 (COH) | Method product_info() is called instead of total_projects() |
| DvM7 (COH) | Method total_projects()is called instead of product_info() |

**Table 10.3(e): Mutants of Product (PM Stands for Product Mutant)**

| Mutants | Fault Description |
|---|---|
| PM1 (ENC) | Product_info() is protected in class product |
| PM2 (COP) | Association product_manufact directs from product to manufacturing |
| PM3 (COH) | Method product_info() is called instead of product_desc() |
| PM4 (COH) | Method product_desc() is called instead of product_info() |

**Table10.3(f) Mutants of Project (PrM Stands for Project Mutant)**

| Mutants | Fault Description |
|---|---|
| PrM1 (COP) | The association development_system directs from project to development |
| PrM2 (COP) | The association project_develop directs from employee to project |
| PrM3 (COH) | Method project_info() is called instead of project_dec() |
| PrM4 (COH) | Method project_dec() is called instead of project_info() |
| PrM5 (COH) | Method project_info() is called instead of team_desc() |
| PrM6 (COH) | Method team_desc() is called instead of project_info() |
| PrM7 (COH) | Method team_desc() is called instead of project_desc() |
| PrM8 (COH) | Method project_desc() is called instead of team_desc() |

**Table10.3(g): Mutants of Sample (SM Stands for Sample Mutant)**

| Mutants | Fault Description |
|---|---|
| SM1 (INH) | Class sample has no product as the base class |

**Table10.3(h): Mutants of Commercial (CM Stands for Commercial Mutant)**

| Mutants | Fault Description |
|---|---|
| CM1 (INH) | Class commercial has no product as the base class |

**Table10.3(i): Mutants of Salaried (SaM Stands for Salaried Mutant)**

| Mutants | Fault Description |
|---|---|
| SaM1 (INH) | Class salaried has no employee as the base class |

**Table10.3(j): Mutants of Contractual (CoM Stands for Contractual Mutant)**

| Mutants | Fault Description |
|---|---|
| CoM1 (INH) | Class contractual has no employee as the base class |

**Table 10.3(k): Results of Mutation Execution**

| Class Name | Killed Mutants | Live Mutants | Mutation Score |
|---|---|---|---|
| Company | CM3, CM4, CM5 | CM1, CM2 | 60% |
| Employee | EM1, EM4, EM5, EM6 | EM2, EM3 | 66.6% |
| Department | DM1, DM2, DM3 | DM4, DM5 | 60% |
| Manufacturing | MM2, MM3, MM4 | MM1 | 100% |
| Development | DvM2, DvM3, DvM4 | DvM1 | 100% |
| Product | PM1, PM3, PM4 | PM2 | 100% |
| Project | PrM3, PrM4 | PrM1, PrM2 | 50% |
| Sample | SM1 | | 100% |
| Commercial | CM1 | | 100% |
| Salaried | SaM1 | | 100% |
| Contractual | CoM1 | | 100% |

Table 10.3(k) shows the result of the mutation execution for each class. In the company mutation, most of COP mutants (CM1, 2) remain live. Adding more test cases could kill these live mutants. The mutation adequacy score of company is 60 percent. The mutants EM2,

EM3 are the live mutants in Employee mutation. The mutation adequacy score of Employee is – 50 percent.

The only live mutant in manufacturing is MM1. MM1 turns out to be an equivalent mutant, so no test case could kill it. The mutation adequacy score of manufacturing is 100 percent, that is, the test set was able to kill all the killable mutants.

The results of mutation execution as shown in table 10.3(k) reveals that even though the test set giving path coverage was able to kill many mutants, some of them remain live, and some more test data needs to be generated to kill these mutants.

## 10.4 CONCLUSION

It is an important testing subject to assess how good generated test sets are, and creation of such an effective and good test set is a critical element of testing process. This paper presents a comprehensive set of mutation operators to test for faults in use of object-oriented features. Object-oriented mutation operators applicable to early in the development life cycle have been classified and evaluated with a case study. Most of the mutants were killed by the sets of tests designed. The most essential theoretical component of a mutation system is that of the mutation operators, thus we have investigated this part of the problem first. However, this theory is not useful until it is followed up by the empirical work.

## KEY TERMS

Mutation, Mutation testing, Mutation analysis, Mutants, Killed mutants, Live mutant, Equivalent mutant, Test adequacy, Non-equivalence mutants, Mutant score, Mutant operators, Object Oriented mutation, Inheritance operators, Encapsulation operator, Coupling operator, Cohesion operator, HLD, LLD,

## REVIEW QUESTIONS

## OBJECTIVE TYPE QUESTIONS

1. Mutation testing is a technique for_______.
    (a) Unit testing
    (b) System Testing
    (c) Integration Testing
    (d) None of the above
2. Mutation analysis is__________ based technique:
    (a) Fault
    (b) Failure

(c) Error
(d) None of the above

3. ___________ operator generates a mutant by changing the structure of class.
(a) Inheritance
(b) Encapsulation
(c) Coupling
(d) Cohesion

4. ____________ operator changes the functional strength of class.
(e) Inheritance
(f) Encapsulation
(g) Coupling
(h) Cohesion

5. ____________ mutant changes the association between classes.
(i) Inheritance
(j) Encapsulation
(k) Coupling
(l) Cohesion

6. Mutant operators are supported to mimic common mistakes made by_______.
(a) Developer
(b) Testers
(c) Both developers and testers
(d) None of the above

## SHORT ANSWER TYPE QUESTIONS

1. 'Developing test often takes sufficient resources'. Justify.
2. List out the major problems test strategies are facing.
3. How to measure the effectiveness of generated test cases? Describe.
4. What do you mean by meta induction?
5. What is mutation testing?
6. Write down the basic approach of testing object oriented software using mutation testing.
7. List out the needs and significance of mutation testing.

8. What are mutants? How mutants are created, and help in testing object oriented software?
9. When a mutant is called killed?
10. What do you mean by equivalent mutants? What the significance of getting equivalent mutants?
11. What is mutant score? How to calculate the mutant score for an object oriented software?
12. How test the adequacy of test set?
13. What is mutation operator? What dos they do?
14. List out the goals of mutation analysis.
15. What do you mean by mutant program? How to generate such a program?
16. How mutant operators are supported to mimic common mistakes made by developers?

## DESCRIPTIVE QUESTIONS

1. Propose few evaluation criteria for test strategies to measure the effectiveness of generated test cases.
2. How mutation analysis is used to assess the quality of test cases? Describe in detail.
3. Generally, software development team views the development as a root-finding procedure. Why? How it helps in testing process?
4. Propose few guidelines for the situation when the test set is inadequate to kill the mutants.
5. Elaborate how mutation testing helps in deciding whether to stop the testing or to enhance the test data set by constructing adequate test sets.
6. Write a paper to test an object oriented software using mutation test in early stage of software development life cycle.
7. Propose a model to develop object oriented mutants early in the development life cycle.

## REFERENCES

1. Philippe Chevally, "Applying Mutation Analysis for Object Oriented Programs using a reflective approach", IEEE 2001 (267-270).
2. B. Beizer, "Software Testing Techniques", Van Nostrand Reinhold, New York, USA, Second Edition, 1996.

3. R. DeMillo, R. Lipton, and F. Sayward, "Hints on Test Data Selection: Help for the Practicing Programmer", Computer, 11 (4): 34-41: April, 1978
4. R. DeMillo, R. Lipton, and F. Sayward, "Program Mutation: A new Approach to Program Testing", Software Testing InfoTech, 1979.
5. R. DeMillo, "Testing adequacy and program mutation", ACM 1989, 355-356.
6. Anna Derezinska, "Object Oriented Mutation to Asses the Quality of Tests", Proceedings of the 19th EUROMICRO Conference "New Waves in System Architecture" (EUROMICRO'03) 2003, IEEE.
7. Y-S Ma, Y-R Kwon, J. Offut, "Inter-Class Mutation Operators for Java", Proc. Of ISSRE '02 IEEE Comp. SOC 2002.
8. S. Ghosh, R. France, C. Braganza, & N. Kawane, "Test Adequacy Assessment for UML Design Model Testing", Proceedings of the 14th International Symposium on Software Reliability Engineering (ISSRE 03), 2003 IEEE
9. J. M. Bieman, S. Ghosh, R. T. Alexxander, "A Techinique for Mutation of java Objects", to appear in Proceedings Automated Software Engineering, 2001 Conference (ASE 2001), Nov. 2001.

## USEFUL LINK

www.computer.org/.../catalog/BP08520& file=preface.xml&xsl=generic.xsl

www.cs.colostate.edu/~trungdt/ uml_testing/uml_testing.html

link.aip.org/link/?JCCEE5/16/66/1

www.businessanalysisbooks.com/0818685204.html

g.oswego.edu/dl/oosdw3/ch6.html

ise.gmu.edu/~ofut/resume-short.html

www.cs.umu.se/~jubo/Meetings/ OOPSLA01/Contributions/RCartwright.html

Chapter 11

# Class Test Complexity

*Courage is not simply one of the virtues, but the form of every virtue at the testing point.*

**C.S. Lewis**

## OBJECTIVES

A successful completion of this chapter will enable the reader for the following:

- To realize the need and importance of object-oriented analysis and design in industrial software development environment.
- To visualize the significance of software metrics.
- To understand the importance of complexity in estimating software quality.
- To list complexity metrics in proposed in literature.
- To understand why LOC is inadequate for object oriented software.
- To enumerate and realize the need and importance of evaluating complexity of software system early in development process.
- To understand class level testing of object oriented software.
- To distinguish object oriented software testing from traditional software testing.
- To understand the role of class-level metrics in testing class complexity.
- To appreciate the need, importance and significance of class level metrics in estimating testing effort.
- To predict probability of code to be reused and quality of class code.
- To appreciate the need of early availability of metrics in software development life cycle.
- To prepare a metric suite for class test complexity.
- To understand that late identification and correction of problems are much more costly than earlier one.
- To implement class-level complexity model to test a module in early stage of software development life cycle.

## 11.1 INTRODUCTION

Object orientation has rapidly become accepted as the preferred paradigm for large-scale system design. Several reasons for this are well known and understood including the following:

- Classes provide an excellent structuring mechanism. They allow a system to be divided into well-defined units, which may then be implemented separately.
- Classes support information hiding. A class can export a purely procedural interface and the internal structure of data may be hidden. This allows the structure to be changed without affecting users of the class, thus simplifying maintenance.
- Object-orientation encourages and supports software reuse. This may be achieved either through the simple reuse of a class in a library, or via inheritance, whereby a new class may be created as an extension of an existing one. In both cases the result is a reduction in the amount of software, which must be written, and, as a result, an improvement in the reliability of the resultant system since previously tested classes may be utilized.

Software testing is often a very costly part of its life cycle. Any technique that improves a software design at an early stage can have highly beneficial impact on the final testing cost and efficiency. Unfortunately the very advantages of object-orientation become potential disadvantages when testing is considered. The structuring of the system as a set of independent classes requires that each of these must be tested and there may be a large number of them. In addition, information hiding, which encourages designers of classes to have purely procedural interfaces, makes it difficult to determine whether the class is working correctly, since the state of internal data may not be accessible via the interface. The result is that the programmer must effectively develop a test program for each class. Each such test program must create an instance of the class being tested and include calls to each of the methods supported by the class. The test program will need to prompt the user for the parameters for these method calls so that various combinations can be exercised. The test program must also display the results of each method call. The result is that the test program may well be more complex and larger than the class being tested.

Object oriented software development techniques may be considered as an advantage to software complexity in software development process. Software metrics have been developed for evaluating and quantifying several aspects of software engineering process. Complexity is directly related to software quality, so representing complexity is important. McCabe identifies a number of important uses for complexity metrics including prediction of critical information about reliability and maintainability of software systems from automatic analysis of source code [or procedural design information], and for feedback during the software project to help to control the design activity.

Moreover, during testing and maintenance, they may provide detailed information about software modules to help to pinpoint areas of potential instability. Several traditional metrics have been proposed in literature, and few of them have been validated. It is evident that the available rational metrics for measuring software like LOC are inadequate for case of object oriented software. These metrics may be applied only to the measurements of the complexity of methods in a class, like McCabe cyclomatic complexity. Several metrics have been proposed for object-oriented software by various researchers and practitioners.

Most of the available metrics may be used in evaluating the complexity of a software system at the final stage of the software production. Much of the research work reveals that maximum efforts have been dedicated with the source code. The determination of complexity for an already written code may be too costly because the later the changes are introduced the more expensive they are. Thus, there appeared to be need for evaluating the complexity of software in early stage of development life cycle without the availability of code. Availability of a suitable and adequate measuring tool at the early stage of development enables early prediction of system complexity thus reducing the cost of making necessary changes.

Many studies have been done on class testing, with the definitions and techniques varying in each work. The class testing used, in general, involves intra-class and inter-class testing. In intra-class testing, each function appearing in the function dependence graph of the class under test is tested by a functional testing technique. The testing starts from the node that has highest fan-in and lowest fan-out, that is, the function which has the most dependents and depends on the fewest functions, then traverses this dependence path and tests the functions on the path. A class-level metrics may be used to identify error prone classes, and hence to estimate testing effort. Further, the class level metrics may be used to estimate the probability of code to be reused, and to improve the quality of class code as well.

## 11.2 CLASS-LEVEL TESTING

The object-oriented paradigm enhances the reliability, maintainability, and reusability of resulting software. It is known as a technique for improving the productivity, quality, and innovation in software development. However, software testing becomes more complex and difficult than that for conventional programming. Object oriented software testing may be at algorithmic level, class level, cluster level, and system level. Class-level testing is the most appropriate way of testing object oriented software. Class-level testing exercises the interactions between the methods in a class or a small cluster of classes. The focus is to verify that the class supports all of the use cases required by its consumers and is robust enough to handle unexpected sequences of methods. There are several reasons to consider generally class-level testing to test object-oriented software including the following:

- Testing at the class level is extremely effective in terms of code coverage. With just a few tests, one can reach 60 percent to 70 percent code coverage.
- Class-level testing reflects the services that the class provides to its consumers. The tests are easy to write and can be used as documentation for the class.
- Class-level testing is the optimal level of testing for finding bugs early in the design process.

There are many possible combinations when methods in a class are invoked. Hence, test cases for object-oriented software at the class level involves not only individual operations but various sequences of operations, which are known formally as 'ground terms'. If two ground terms are equivalent according to the specification, but their implemented method sequences generate observationally non-equivalent objects, then there is an error in the implementation. Using this concept, pairs of equivalent ground terms should be selected as test cases for any given class. This recommendation cannot be directly applied in practice, however, because the set of all equivalent pairs is infinite in most circumstances.

## 11.3 CLASS METRICS

No doubt, almost all research in the field of object-oriented metrics has been directed towards association with source code a complexity number. However class complexity evaluation after a class code had been already written may be too expensive. One of the widely accepted metric suite by Chidamber and Kemerer, includes following six object-oriented metrics based upon measurement theory.

- Number of Children (NOC)
- Depth of Inheritance Tree (DIT)
- Weighted Methods Per Class (WMC)
- Response for a Class (RFC)
- Coupling Between Objects (CBO)
- Lack of Cohesion in Methods (LCOM)

**Number of Children (NOC):** The NOC metric is defined as the count of immediate subclasses. With greater NOC the likelihood of improper abstraction of parent class is greater. The NOC may also give an idea of the potential influence a class has on the design. If a class has a large NOC value, it may require more testing of the methods in that class.

**Depth of Inheritance Tree (DIT):** DIT is the length of the maximal path to the root of the class hierarchy. It can be observed that the deeper is a class in the hierarchy, the greater can be the number of inherited methods making it more difficult to predict its behavior so more effort is needed in testing this class. Deeper trees constitute greater design complexity, since more methods and classes are involved.

**Weighted Methods per Class (WMC):** WMC is the count of methods in a class. The number of methods and its complexity is a predictor of how much time and effort is required to develop and maintain the class. Classes with large number of methods are likely to be more application specific, limiting the possibility of reuse. The more methods defined in a class the greater is its possible impact on children, since children inherit all defined methods.

**Response for a Class (RFC):** The RFC metric is the cardinality of the set of all methods that can be potentially executed in the response to the arrival of a message to an object. RFC is a measure of the potential communication between the class and other classes. If a large number of methods can be invoked in response to a message, the testing and debugging of the class becomes more complicated since it requires a greater level of understanding required on the part of the tester. The larger the number of methods that can be invoked from a class, the greater the complexity of the class. A worst case value for possible responses will assist in appropriate allocation of testing time.

**Coupling between Objects (CBO):** The CBO metric is defined as the number of couples with other classes (where calling a method or using instance variable from another class constitutes coupling). The more independent a class is, the easier it is to reuse it in another application. To improve modularity and promote encapsulation inter class coupling should be small. The larger the number of couples, the higher is the sensitivity to changes in other parts of the design and therefore maintenance is more difficult. CBO metric is useful to determine how complex the testing of various parts of a design is likely to be. The higher the CBO is, the more rigorous the testing needs to be. CBO depends on the manner in which methods are designed and not on the functionality of the class.

**Lack of Cohesion in Methods (LCOM):** The LCOM metric is defined as a count of the method pairs that do not have common instance variable minus the count of method pairs that do. The larger the number of similar methods, the more cohesive is the class. If none of the methods of a class display any instance behavior, i.e. do not use any instance variables, they have no similarity and the LCOM for the class will be zero. Cohesiveness of methods within a class is desirable, since it promotes encapsulation. Lack of cohesion implies classes should be probably split into two or more subclasses. Any measure of disparity of methods helps identifying flaws in the design of classes. Low cohesion increases complexity, thereby increasing the likelihood of errors during the development process.

Lorenz and Kidd (L&K) proposed another metric suite including the following ten metrics. Unlike the Chidamber-Kemerer (C&K) metrics, most of the L&K metrics are direct metrics, and include more directly countable measures, e.g., the Number of Methods (NM) metric, and the Number of Variables (NV) metric. Although relatively simple to collect, doubt can be cast on the usefulness of such metrics because they give only a limited insight into the architecture of the system under investigation. For each of the metrics considered, L&K offered some justification for the existence of that metric which is included here. The ten metrics can be summarized as:

**Number of Public Methods (PM):** The PM metric simply counts the number of public methods in a class. This metric is useful as an aid to estimating the amount of work to develop a class or subsystem.

**Number of Methods (NM):** NM is the total number of methods in a class counts all public, private and protected methods defined. This metric may be used as a useful indication of the classes, which may be trying to do too much work themselves; i.e., they provide too much functionality.

**Number of Public Variables per class (NPV):** The NPV metric counts the number of public variables in a class. L&K consider the number of variables in a class to be one measure of its size. The fact that one class has more public variables than another might imply that the class has more relationships with other objects and, as such, is more likely to be a *key class,* that is, a central point of co-ordination of objects within the system.

**Number of Variables per class (NV):** NV metric counts the total number of variables in a class. The total number of variables metric includes public, private and protected variables. The ratio of private and protected variables to total number of variables indicates the effort required by that class in providing information to other classes. Private and protected variables are therefore viewed merely as data to service the methods in the class.

**Number of Methods Inherited by a subclass (NMI):** This metric measures the number of methods inherited by a subclass. No mention is made as to whether that inheritance is public or private. In a language such as C++, we have to consider the possibility that the inheritance may be private. Then, any classes using methods from a subclass would not necessarily have access to all of the inherited methods.

**Number of Methods Overridden by a subclass (NMO):** A subclass is allowed to re-define or override a method in its superclasses with the same name as a method in one of its superclasses. According to L&K, a large number of overridden methods indicate a design problem, indicating that those methods were overridden as a design afterthought. They suggest that a subclass should really be a specialization of its superclasses, resulting in new unique names for methods.

**Number of Methods Added by a subclass (NMA):** According to L&K, the normal expectation for a subclass is that it will further specialize (or add) methods to the superclass object. A method is defined as an added method in a subclass if there is no method of the same name in any of its superclasses.

**Average Method Size (AMS):** The average method size is calculated as the number of non-comment, nonblank source lines (NCSL) in the class, divided by the number of its methods. AMS is clearly a size metric, and would be useful for spotting outliers, that is, abnormally large methods.

**Number of times a Class is Reused (NCR):** The definition of NCR given by L&K is somewhat ambiguous. We assume the metric is intended to count the number of times a class is referenced (that is, reused) by other classes. In this sense, we could view reuse in a similar way to coupling. We could then consider NCR as a measure of the extent of inter-class communication, and in this respect, a high value for NCR as undesirable.

**Number of Friends of a class (NF):** This metric measure, for each class, the number of friends of that class. Friends allow encapsulation to be violated, and as such should be used with care. A high number of friends within a class could indicate a potential design flaw, an oversight in design, which has filtered through to the coding stage; we note in passing that friends are a concept specific to the C++ language. NF is a measure of class coupling, since friends may rely on a particular class (or classes) to operate properly.

Another set of six metrics developed by Abreu was intended to be design metrics. Abreu worked in the area of object oriented design metrics and has proposed a set of object oriented metrics MOOD comprising of MHF, AHF, MIF, AIF, PF and CF. The research was carried out by the Software Engineering Group at INESC in cooperation with the Lisbon Technical University to fulfill two goals for the set of metrics development. The first goal was to improve the object oriented design process to achieve better maintainability. The second goal was to improve the object oriented estimation process to achieve better resource allocation. The six metrics have been summarized and presented in the following section.

**Polymorphism Factor (PF):** This metric is based on the number of overriding methods in a class as a ratio of the total possible number of overridden methods. Polymorphism arises from inheritance, and Abreu claims that in some cases, overriding methods reduce complexity, so increasing understandability and maintainability.

**Coupling Factor (CF):** This metric counts the number of inter-class communications. There is a similarity here with the NCR metric of L&K. Abreu views coupling as increasing complexity, reducing both encapsulation and potential reuse and limiting understandability and maintainability.

**Method Hiding Factor (MHF):** This metric is the ratio of hidden (private or protected) methods to total methods. As such, MHF is proposed as a measure of encapsulation.

**Attribute Hiding Factor (AHF):** This metric is the ratio of hidden (private or protected) attributes to total attributes. AHF is also proposed as a measure of encapsulation.

**Method Inheritance Factor (MIF):** This metric is a count of the number of inherited methods as a ratio of total methods. There is a similarity here with the NCR metric of L&K. Abreu proposes MIF as a measure of inheritance, and consequently as a means of expressing the level of reuse in a system. It could also claim to be an aid to assessment of testing needed.

**Attribute Inheritance Factor (AIF):** This metric count the number of inherited attributes as a ratio of total attributes. Just as for the MIF, Abreu proposes AIF as a means of expressing the level of reuse in a system. It is claimed, however, that too much reuse causes deterioration in understandability and testability.

## 11.4 OBJECT ORIENTED METRICS - CURRENT SCENARIO

A range of object oriented metrics have been developed that may be used to evaluate and predict the complexity and at last the quality of software. Object oriented programming metrics is an aspect to be considered to provide the dependable guidelines for object oriented analysis and design software development life cycle. Object oriented metrics, which may be applied to analyze source code as an indicator of quality attributes. Numerous object oriented software metrics related to software complexity and quality assurance have been proposed in the past.

A testing engineer may choose from the large set of object-oriented metrics. But the problem a tester has to face is the selection of an appropriate set of metrics for his measurement because there is no a single metric which may incorporate all the aspects of object oriented design. Thereby, the question posed is not the lack of metrics but the selection of those metrics, which meet the specific needs of each software project. A number of object-oriented metrics exploit the knowledge gained from metrics used in structured programming and adjust such measurements so as to satisfy the needs of object-oriented programming.

Apropos to the forgoing references it is evident that early availability and use of metrics in SDLC is a key factor to a successful management of software development, and a leading factor for reducing the class complexity test effort. Moreover, such a practice may prove to be highly fruitful for many reasons including following:

- Early detection of problems in the artifacts produced in the initial phases of the lifecycle (specification and design documents);
- Reduction of the cost of change—late identification and correction of problems are much more costlier than earlier ones;
- Better software quality monitoring from the early phases of the life-cycle;
- Quantitative comparison of techniques and empirical refinement of the processes to which they are applied;
- More accurate planning of resource allocation based upon the predicted error proneness of the system and its constituent parts.

A comprehensive review on metrics suggested by various researchers and practitioners, indicates that the object oriented metric ultimately rely on *class*, so it appears to be good to develop a suit of class-based metric which may be used to predict the class test complexity incorporates all aspects of object oriented design, that is it takes into account encapsulation, inheritance, coupling as well as cohesion. Thereby it may possibly indicate in whole the complexity of the entire class based on all these factors.

## 11.5 METRICS SUITE

Several research works in the object oriented metrics arena were produced in recent years. However, widespread adaptation of object oriented metrics in numerous application domains should only take place if the metrics may be shown to be theoretically valid, in the sense that they accurately measure the attributes of software for which they were designed to measure and have also been validated empirically. But there is a general agreement among experts that metrics have not been even validated theoretically. Most of the metric is accepted by practitioners on 'heavy usages and popularity' and by academic experts on empirical (post development) validation. In such a scenario many of the available metrics may not be used properly and have been discarded. Reasons may be that these could not find a place among practitioners or not found to be valid by experts on empirical findings. Now the question is 'Why such metrics which could not become acceptable were developed?' Leading to all efforts in development going vain.

This led to the definition of four new metrics: Encapsulation Metrics (ENM), Reuse Metrics (REM), Coupling Metrics (CPM) and Cohesion Metrics (CHM), which could be calculated from design information only. A suite of object-oriented metrics, covering all the design attributes, has been listed and described in table 11.4

**Table 11.4: Proposed Metrics**

| Metrics | Name | Description |
|---|---|---|
| **ENM** | Encapsulation Metric | This metric count of all the methods defined in a class |
| **REM** | Reuse Metric | This metric count of the number of class hierarchies in the design. |
| **CPM** | Coupling Metric | This metric count of the different number of classes that a class is directly related to. |
| **COM** | Cohesion Metric | This metric computes the relatedness among methods of a class based upon the parameter list of the methods [computed as LCOM, 1993 Li and Henry version] |

Thereby, the suite gives way to fix component weightage using linear regression method and hence fitting the regression line for computation of a key attribute to quality, complexity.

The extensive review of object oriented development literatures indicate that encapsulation, inheritance and coupling have been generally viewed to promote the complexity. Based upon the design property to quality attribute relationship, the relative significance of individual design properties that influence complexity is weighted proportionally. A multiple linear regression (MR) has been used to get the coefficients. The multiple linear regressions establish a relationship between dependent variables and multiple independent variables. The MR equation takes the form:

$$y = \beta_0 + \beta_1 x_1 + \beta_2 x_2 + \ldots\ldots\ldots\ldots \beta_k x_k \qquad (1)$$

Where

- The xs are the independent (predictor or regressor) variables.
- Y is the dependent (response) variable.
- β (Regression coefficient) is the average amount of dependent increase when the independents are held constant.

Complexity of a class depend upon one or more number of design attributes, component-wise effect may be speculated and respective component weightage (CW) may be fixed using regression equation. Thereby, the CWs of individual design attributes have been calculated in terms of regression coefficient β. Three medium size C++ projects were used to fit the regression line by acquiring real data from commercial projects of Software Company based in Delhi, India. Name of the projects and actual source data is being concealed, as per wishes of company's management. We have assured authenticity of source data, to the best possible extent. These projects include the number of classes (10-20). The projects were independently completed over a period four months. Using these data, the CW coefficient calculated for encapsulation, inheritance, coupling and cohesion to show the complexity relationship with design properties (encapsulation, inheritance, coupling and cohesion). Equation 2 summarizes the computational formula for Complexity with the component weightage.

$$\text{Complexity}=0.16 * \text{Encapsulation} + 2.78 * \text{Inheritance} + 1.57 * \text{Coupling} \quad (2)$$

## 11.6 METRIC SUITE CALCULATION ON CLASS DIAGRAM

The following four metrics proposed in this paper may be easily calculated from the class diagram. Encapsulation Metrics (ENM) can be calculated as soon as class operations are identified. Reuse Metrics (REM) can be calculated by counting of the number of class hierarchies in the design shown in figure 10.3. Here, to calculate the proposed Coupling Metrics (CPM) and Cohesion Metrics (COM), the source code of methods are not required.

**Table 11.6: Metrics Calculation**

| Class Name | ENM | REM | COP | COM |
|---|---|---|---|---|
| Company | 3 | 0 | 2 | 3 |
| Employee | 3 | 0 | 3 | 1 |
| Department | 2 | 0 | 2 | 1 |
| Manufacturing | 3 | 1 | 1 | 1 |
| Development | 4 | 1 | 1 | 2 |
| Product | 2 | 0 | 1 | 2 |
| Sample | 3 | 1 | 0 | 2 |
| Commercial | 2 | 1 | 0 | 1 |
| Project | 3 | 0 | 2 | 1 |
| Salaried | 3 | 1 | 0 | 1 |
| Contractual | 3 | 1 | 0 | 1 |

The detail-calculated values of all the four metric suits are shown in table 11.6 for the figure10.3.

## 11.7 AN EXPERIMENTAL VALIDATION

This section assesses how well the proposed metrics suit is able to predict the 'Class Test Complexity' of object-oriented software. The internal characteristic of a design varies significantly with the objective and domain of the design. These characteristics influence the complexity and other quality attributes and, therefore, the overall quality. A medium size C++ project was used in the validation of metrics suit. This project was selected because several designs of the project that addresses the same set of requirements were easily available. The size of these projects, reflected by the number of classes (10-20), made them an ideal candidate. The projects were independently completed over a period of months.

A suite of six projects was developed by different individuals for the study. The six projects of the validation suite were also evaluated using these metrics. The metrics values calculated for the four metrics used to assess the design properties are shown in table 11.7(a). The values for complexity are computed using the equations 2.0 and shown in table 11.7(b). These complexity scores are used to assign the Complexity Benchmark (BC) to all the six projects.

**Table 11.7(a): Design Property Metrics Value**

| **Metrics** | **Projects** | | | | | |
|---|---|---|---|---|---|---|
| | 1 | 2 | 3 | 4 | 5 | 6 |
| **Encapsulation** | 3.3.6 | 2.88 | 3.91 | 4.10 | 4.22 | 2.86 |
| **Inheritance** | .36 | .33 | .58 | .42 | .55 | .26 |
| **Coupling** | 1.81 | 1.88 | 2.08 | 2.52 | 2.72 | 2.26 |
| **Cohesion** | 1.27 | 1.33 | 1.58 | 1.89 | 1.33 | 1.26 |

A group of ten independent evaluators was assigned to study the quality of the six projects in the validation suite. All the evaluators had 8 to 12 years of experience in commercial software development, had knowledge of the object oriented paradigm, and had developed software using C++. The study was done over a period of month. All the participants analyzed each project's design and assigned the scores on a ordinal scale to the Complexity by using their own traditional tool. These scores for ll five high-level quality attributes were summed separately to give the evaluator's Complexity Benchmark (BC) for the project. The projects were ranked 1 through 6 based on decreasing complexity scores by each evaluator as shown in table 11.7(c).

**Table 11.7(b): Design Quality Attribute Indices**

| **Quality Factors** | 1 | 2 | 3 | 4 | 5 | 6 |
|---|---|---|---|---|---|---|
| **Complexity** | 20.18 | 15.14 | 30.47 | -25.23 | 50.64 | 24.73 |
| **BC** | 3 | 2 | 5 | 1 | 6 | 4 |

The ranking of validation suite projects indicates a closer agreement between the assessments done by the evaluators. There was a very strong agreement between the evaluators in the assessment of project 2 and 4. Also projects 3 and 5 are very closely ranked by at least 5-6 evaluators.

**Table 11.7(c): Design Quality Attribute Indices**

| **Project No.** | **BC** | **Evaluators Ranking** | | | | | | | | | |
|---|---|---|---|---|---|---|---|---|---|---|---|
| | | 1 | 2 | 3 | 4 | 5 | 6 | 7 | 8 | 9 | 10 |
| 1 | 3 | 3 | 4 | 3 | 3 | 4 | 4 | 5 | 3 | 4 | 4 |
| 2 | 2 | 2 | 3 | 2 | 2 | 2 | 3 | 3 | 2 | 2 | 3 |
| 3 | 5 | 6 | 5 | 4 | 5 | 6 | 6 | 5 | 5 | 5 | 6 |
| 4 | 1 | 1 | 2 | 3 | 2 | 1 | 2 | 2 | 2 | 1 | 1 |
| 5 | 6 | 5 | 5 | 5 | 6 | 5 | 6 | 6 | 6 | 6 | 6 |
| 6 | 4 | 5 | 5 | 4 | 3 | 4 | 4 | 4 | 4 | 4 | 4 |

## 11.8 STATISTICAL ANALYSIS

The Spearman's rank correlation coefficient, $r_s$ was used to test the significance of correlation between metric design-based Complexity assessment and the evaluator's implementation based assessment of the project. It provides a nonparametric significance test that works well with ranked data without precise proportional scaling and can be used to detect relationships other than linear one. For the level of significance of $\alpha$=95%, the threshold value calculated for n=6 projects was$\pm$. 817.

For two independent evaluations X1 and X2 of n items that are to be correlated, the values of X1 and X2 are ranked from 1 to 'n' according to their relative size within the evaluations. For each X1 and X2 pair in relative rank, differences in ranks'd' is computed. The sum of all $d^2$ s denoted by $\sum d^2$ is used to compute $r_s$ using formula:

$$r_s = 1 - \frac{6\sum d^2}{n(n^2-1)} \qquad -1.0 \leq r_s \leq +1.0 \qquad (3)$$

$r_s$= Coefficient of Rank Correlation
n= No. of Paired Observation
$\sum$= Notations meaning 'the sum'
d= difference between the ranks for each pair of observations.

## 11.9 CONTEXTUAL INTERPRETATION

The correlation values between metric suits determined complexity rankings of the six projects and ten evaluators assessments of overall quality of these projects based on design and implementation shown in table 11.7(c), has been listed in table 11.9. Pairs of Metric Suits – Evaluator with Correlation values $r_s$ above (±. 817) are checked in table 11.9. It is evident from the correlation values shown in table 11.8 that the complexity indicators evaluated using proposed metrics suits are highly correlated with the measurements done by human evaluators.

**Table 11.9: Metric Suits and Evaluators Ranking Correlation's of Projects**

| | **1** | **2** | **3** | **4** | **5** | **6** | **7** | **8** | **9** | **10** |
|---|---|---|---|---|---|---|---|---|---|---|
| $\Sigma d^2$ | 3 | 5 | 6 | 2 | 4 | 4 | 6 | 1 | 1 | 3 |
| $r_s$ | .92 | .85 | .82 | .94 | .89 | .89 | .82 | .97 | .97 | .92 |
| $r_s$> .817 | √ | √ | √ | √ | √ | √ | √ | √ | √ | √ |

## 11.10 CONCLUSION

Numerous metrics have been proposed in literature. Most of the proposed metrics may be used in evaluating the complexity of a software system at the final stage of the software production. Much of the research work reveals that maximum efforts have been dedicated with the source code. It has been revealed from the literature survey that a class-level metrics may be used to identify error prone classes, and hence to estimate testing effort. Further, it may also be used to estimate the probability of code to be reused, and to improve the quality of class code as well.

The proposed metrics suits for the assessment of complexity in object-oriented design have been validated using structural and functional information from object oriented software. The metric suits' ability to estimate overall complexity from design information has also been demonstrated using several functionally equivalent projects where the overall complexity estimate computed by model had statistically significant correlation with the assessment of overall project complexity characteristics determined by independent evaluators. It is apparent that metrics suit may be used effectively in monitoring the complexity and hence overall quality of software in design phase. However, in order to generalize and standardize, experiments may be needed at a large scale.

## KEY TERMS

Object oriented metrics, Complexity metrics, Class-level metrics, Class Test, Software Quality, Quality factors, Quality Metrics, Design Metrics, Encapsulation metrics, Reuse metrics, Coupling metrics, Cohesion metric, Component weightage, Class test complexity, Complexity benchmarking, Object Oriented design property,

## REVIEW QUESTIONS

### OBJECTIVE TYPE QUESTIONS

1. ___________ metric may be used estimate the probability of the code to be reused.
   - (a) Class level
   - (b) Code level
   - (c) Design level
   - (d) None of the above
2. Error prone classes may be identified by_______.
   - (a) Class level
   - (b) Code level
   - (c) Design level
   - (d) None of the above
3. LOC is inadequate for_______.
   - (a) Traditional software
   - (b) Object oriented software
   - (c) Both traditional and object oriented software
   - (d) None of the above
4. Class complexity test effort may be reduced by _______.
   - (a) Early availability and use of metrics in SDLC
   - (b) Availability and use of metrics in code phase of SDLC
   - (c) Availability and use of metrics in code or design phase of SDLC
   - (d) None of the above
5. Object oriented metrics rely on_______.
   - (a) Methods
   - (b) Class
   - (c) Both on class and method
   - (d) None of the above

### SHORT ANSWER TYPE QUESTIONS

1. How complexity is related with software quality?

2. Define software instability.
3. List out the traditional metrics proposed in the literature.
4. Why complexity determination for already written code is too much difficult?
5. What is the advantage of evaluating software complexity in early stage of development life cycle?
6. Explore the possibility of evaluating complexity without the availability of source code.
7. How can one identify error prone classes?
8. What is the role of software metrics in estimating testing efforts?
9. Why late identification and correction of problems are much costlier than earlier one?
10. How software can better be monitored from early phases of the life cycle?
11. How a suite of class based metric helps in estimating object oriented software complexity in early stage?

## DESCRIPTIVE QUESTIONS

1. Write down the set of guideline a tester can follow during selection of metrics to estimate complexity of software.
2. Propose a single metric which may incorporate all the aspect of object oriented design.
3. Validate the proposed metrics in previous question. Implement the metric to estimate the complexity of object oriented software.
4. Find out the commonly accepted set of the factors for software complexity.
5. 'There is not the lack of metrics in the literature, but the selection of appropriate metric which meet the specific needs of each software project'. Justify.
6. 'An early availability and use of metrics in SDLC is a key factor to a successful management of software development.' Justify.

## REFERENCES

1. Llona Nbluemke, "Object oriented metrics useful in the prediction of class testing complexity", IEEE 2001.
2. I. Bashir, A. L. Goel, "Metrics: A guide for OO Software Testing", Proceedings of Quality Week, SanFrancisco, 1997.
3. I. Bluemake, "Object Oriented Metrics for CASE product", Proceedings of $20^{th}$ ASU Conference OO Metrics & Simulation, University of Malta, Editor J. Sklenar, Malta, pp. 55-64, Sept. 2000

4. McCabe, T. J., "A Complexity Measure", IEEE Transactions on Software Engineering, Vol. 2, No.4, December 1976, pp. 308-320.

5. Bansiya Jagdish & Devis Carl, "Automated Metrics and Object Oriented Development", Dr. Dobb's Journal December 1997.

   http://www.ddj.com/documents/s=934/ddj9712d/9712d.htm

6. Bansiya Jagdish, "A Hierarchical Model for object- oriented Design Quality Assessment", IEEE Transaction on software engineering, Vol.28, No.1, January 2002.

7. Dumke, Reiner R., "A Measurements framework for Object- Oriented Software Development", submitted to Annals of Software Engineering , Vol.1, 1995

8. Henderson Sellers, B., "Identifying internal and external characteristics of classes likely to be useful as structural complexity metrics", Proceedings of 1994 intern. Conf. On Object Oriented Information Systems OOIS'94, London, December 1995, Springer-Verlag, pp. 227-230, London, 1995.

9. Campanai M. and Nesi P., "Supporting Object Oriented Design with Metrics", Proceedings of TOOLS Europe' 94, France, 1994.

10. Cant S. N., B. Henderson Sellers, and Jeffery D. R., "Application of cognitive complexity metrics to object- oriented program", Journal of Object- Oriented Programming, pp. 52-63, July- August 1994.

11. Hopkins, Trevor P., "Complexity Metrics for Quality Assessment of Object- Oriented Design", SQM'94, Edinburgh, July 1994, Proceedings published as Software Quality Management II, vol. 2: Building Quality into Software, pp. 467-481, Computational Mechanics Press, 1994.

12. Abreu, Brito F. and Carapuca, Rogerio, "Candidate Metrics for Object- Oriented Software within a Taxonomy Framework", Proceedings of AQUIS'93 (Achieving Quality In Software), Venice, Italy, October 1993: selected for reprint in the Journals of Systems and Software, Vol. 23 (I), pp. 87- 96, July 1994.

13. Littlefair Tim, "An Investigation into the use of software code metrics in the industrial software development environment", A Thesis Submitted in Partial Fulfillment of the requirements for the award of Ph. D, 18th June 2001.

14. Khan R. A. & Mustafa K., "A Process for Assessing Object Oriented Software System", Developer IQ, Software Technology Magazine, Techmedia, India, Vol. 4, No. 4 April, 2004.

15. Xenos M., Stavrinoudis D., Zikouli K. and Christodoulakis D. "Object Oriented Metrics- A Survey", Proceeding of the FESMA 2000, Federation of European Software Measurement Associations, Madrid, Spain, 2000

16. Chidamber, S. R. and C. F. Kemerer, "Towards a metrics suite for Object- Oriented Design", Proceeding: OOPSLA '91, July 1991, pp. 197-211.

17. Li., W. and Henry, "Maintenance metrics for the object-oriented paradigm", Proceeding Of The First International Software Metrics Symposium, May 1993, pp.-52-60.

18. Hitz, M. and Montazeria. B., "Chidamber and Kemerer's Metrics Suite: a measurement theory perspective", IEEE Transactions on software Engineering, 22(4), April 1996, pp. 267-271.
19. Schneidewind N. F., "Methodology for validating software metrics", IEEE Software Engineering, vol. 18, no. 5, pp. 410-422, 1992.
20. Khan R. A., Mustafa K., & Yadav S., "Quality Assurance of Object Oriented Code in Design Phase", Proceedings of 4th Annual International Software Testing Conference, India Feb 20-21, 2004.

## USEFUL LINKS

www.iro.umontreal.ca/~sahraouh/qaoose01/Reissing.pdf

turing.une.edu.au/~comp284/ Lectures/Lecture_20/lecture/node1.html

www.cs.virginia.edu/~horton/ cs494/s05/slides/UnitTesting-lecture-s05.ppt

courses.cs.tamu.edu/cpsc431/lively/Metrics.ppt

www.pitt.edu/~ckemerer/CK%20research%20papers/ TSE0024-0205R1-FinalPaper.doc

www2.umassd.edu/SWComprehension/ compdocs/durham/2-96.pdf

www.bluej.org/papers/1998-11-OO-testing.pdf

hissa.nist.gov/HHRFdata/ Artifacts/ITLdoc/235/chapter8.htm

www.cs.uah.edu/~letzkorn/joop2.pdf

www.cs.clemson.edu/~johnmc/joop/col3/column3.html

# INDEX